THE HISTORY OF
UNDERWATER EXPLORATION

BY ROBERT F. MARX

DOVER PUBLICATIONS, INC., NEW YORK

*This book is dedicated to
John Gaffney, one of the great
pioneers of the underwater world*

Published in Canada by General Publishing Company, Ltd., 30 Lesmill Road, Don Mills, Toronto, Ontario.

Published in the United Kingdom by Constable and Company, Ltd., 3 The Lanchesters, 162–164 Fulham Palace Road, London W6 9ER.

This Dover edition, first published in 1990, is a revised republication of the work originally published in 1978 by Van Nostrand Reinhold Company, New York, under the title *Into the Deep: The History of Man's Underwater Exploration*. The present edition features a new selection of pictures in a totally new layout, with corresponding captions.

Manufactured in the United States of America
Dover Publications, Inc., 31 East 2nd Street, Mineola, N.Y. 11501

Library of Congress Cataloging-in-Publication Data

Marx, Robert F.
    The history of underwater exploration / Robert F. Marx.
        p.   cm.
    Rev. ed. of: Into the deep. c1978.
    Includes bibliographical reference (p.   ) and index.
    ISBN 0-486-26487-4 (pbk.)
        1. Diving, Submarine—History. 2. Underwater exploration—History.
    I. Marx, Robert F., 1933– Into the deep. II. Title.
    VM977.M37   1990
    627'.72'09—dc20   90-3524
                                                                      CIP

CONTENTS

INTRODUCTION

"Under Water Men Shall Walk, Shall Ride, Shall Sleep and Talk."
Martha (Mother) Shipton (1488–1561)

In an age when people's imaginations are caught by the possibility of breakthroughs in the exploration of space, it is easy to forget that there are still places on this planet that have not been fully explored yet. True, we have learned much about the earth's environment, but much remains to be learned, particularly about the more than two-thirds of the earth's surface that is covered with water. In spite of all the advances in science and technology, the deepest parts of the seas and oceans of this planet remain a mystery to man.

Statisticians predict that by the year 2000, the earth's present population of about 3.5 billion will have doubled, to around seven billion. Yet even with our present population, over two billion people are undernourished. Obviously, therefore, man must augment his age-old methods of fishing the seas and searching the seafloors for food with scientific methods of farming the oceans. We must also increase our efforts to mine the seas' mineral wealth, as well as find new and better ways to coexist with the seas, developing the potential of both.

There is currently a growing conviction among scientists and businessmen that the next major resource to be tapped is the sea. With the earth's population expanding geometrically and thus accelerating the depletion rate of the natural and mineral resources on land, it is inevitable that we will look to the oceans for the means of sustaining ourselves. To accomplish this, however, we must first learn to live and work underwater.

1

For centuries men have struggled to penetrate the ocean depths against barriers nearly as formidable as those faced by today's astronauts. In water, as in space, man enters an environment he cannot adjust to without help. The most obvious help he needs is in breathing. The amount of time he can hold his breath underwater varies according to the capacity of his lungs, his metabolism, and his general physical condition. It also varies with the amount of energy he must expend. Sitting at the bottom of a pool, he can hold his breath much longer than if he were chasing fish with a spear or repairing a ship's bottom. But if he tries to hold his breath too long, he risks losing consciousness because of a lack of oxygen; when that happens, water floods his lungs and he drowns.

Another problem we face underwater is the pressure. The pressure of the air around us is only 14.7 pounds per square inch, so the human body, which is accustomed to it, doesn't even feel it. But water is much denser than air, and its pressure increases by "one atmosphere" (the equivalent of atmospheric pressure, or 14.7 lbs./sq. in.) for every 33 feet we descend into the water. This means, for example, that the pressure at 66 feet is 44.1 pounds per square inch, or three times that of the atmosphere. Our lungs can adjust to three atmospheres of pressure, even to four if they are exceptionally strong. So can our ears if we hold our noses and force air through the Eustachian tube to equalize the pressure. But there is always the danger of making a mistake and having the pressure cause our eardrums to rupture. We soon reach a depth at which adjustment is impossible, for now the body itself will be crushed by pressure of the water.

When we think of such incredible forces to be overcome, we are even more awed by the fact that since the beginning of recorded history and beyond, men have been diving into the sea with nothing more than their human strength and courage. They did so for various reasons: to hunt for food, to surprise their enemies in warfare, to search for treasure, to satisfy a love of adventure, to match their strength and skill against that of others, and eventually to further scientific knowledge. These reasons have always existed, but thanks to man's advancing technology, the difficulties involved have been diminished. This technology has reached the stage where divers are equipped to penetrate to depths that were unthinkable as recently as 50 years ago. The struggle won't be over, though, until the deepest parts of the oceans have been charted as well as the rest of the sea.

It probably isn't mere coincidence that the ratio of salt to fresh water in human blood is the same as that of the sea. While we don't know the origins of life, there is considerable evidence which suggests that it began in the oceans. Fossil remains of organisms that lived almost three billion years ago have been discovered in Barberton, South Africa. Until about 20 years ago, it was thought that life on earth originated early in the Cambrian period, approximately 550 million years ago; but recent research indicates that it is more than 27 billion years old.

The earliest known forms of life were one-celled organisms closely related to bacteria, which lived in the sea. Eventually these cells evolved into other species of marine life, many of which are still in existence. Some—like the plesiosaurus sea reptiles, which lived during the time of the giant dinosaurs and which may have been as much as 300 feet long—disappeared a long time ago. Fossilized shark remains dating back 400 million years have been found in many areas of the United States that were once underwater. Some of these creatures were over 100 feet long. They, too, have disappeared. Otherwise, diving would not enjoy the popularity it does today!

Many ancient peoples were terrified of the shark (or "sea dog," as they called it). Among the sea's inhabitants, human attributes were divided between sharks and merfolks. Merfolks were graceful creatures which exhibited an imaginative ability to play among themselves, whereas sharks were viewed with horror, not just because they attacked people, but because of their peculiar and sinister relationship to man. Savages in West Africa, who claimed that sharks could take human form and human mates, would sacrifice children to shark gods. In the shark temples of the Sandwich Islands, priests rubbed their skin with rough objects to make it scaly like that of the sharks they venerated, and in northern Europe during the Middle Ages, offerings of coins were thrown into the sea when sharks were spotted at sea. Until well into this century, it was a capital offense in many parts of the South Pacific to kill a shark. Early mariners wore gold amulets containing sharks' teeth, in the belief that this would protect them from drowning.

Myths about mermaids date back to the time when the Phoenicians dominated the seas. They are believed to have originated the legend to frighten off sailors from competing maritime countries. According to Ernest Buffon, a Frenchman, writing in 1481:

> The siren is a monster of the sea. The upper part of its body is that of a virgin whilst the lower part is that of a fish. It sings so sweetly that men who hear its songs are charmed to sleep, to wit, the mariners. And in this way more than one youth sailing the sea, having seen the siren, has been consumed with love for her and desired to have her fleshy company for her beauty and for the sweetness of her song. Aware of this the siren invites the youth to approach, saying: "Fair youth, it seems you love me. If this be so, approach without fear and I will grant you all your desires." The youth, mad with love and desire, then springs from his ship into the sea to embrace the siren and hold her in his arms. But she swims away and leaves him there to drown.

For thousands of years men have regarded the sea as dangerous and menacing but always fascinating. What the sea was to the seamen of Crete, it still is today, both to the poor fisherman trying to wrest a living from it and the rich man who sails on it for pleasure. In its darker moods, the sea seems to be a symbol of eternity to some; but for centuries it was more. Men shrank from contemplating it as they might shrink from thoughts of eternity. A voyage far out to sea was as forbidding as a voyage of the soul into the shadows beyond the grave. In fact, the ocean currents that flow around the world were once believed to be a great river flowing by the throne of God.

The ancient Greeks revered Oceanus as the "father of all creation" who lived at the westernmost edge of the world amid a court of sea monsters. Oceanus would sit on his throne, stroking his long white beard with red, crab-like claws as he watched with all-seeing eyes man's struggle to master the sea. Three of the 12 gods enthroned by the Greeks on Mount Olympus lived under the sea: crookleg Hephaestus, the deity of invention and engineering; Aphrodite, the goddess of love; and blustering Poseidon, minister of the oceans. Poseidon shared the sovereignty of the universe with the landsmen, Zeus and Hades. According to Homer, Poseidon won the more desirable Nereide, Amphitrite, by sending his fastest porpoise to catch her. Their son, Triton, was well equipped for diving. From the waist up, he was a man, but from there down, he was outfitted with flukes—the original merman. Triton carried a conch trumpet which he blew sweetly to quiet stormy seas and loudly to bring a gale.

Poseidon, called Neptune by the Romans, was the chief god of the sea. He lived under the Aegean Sea in a magnificent palace with walls of mother-of-pearl, hangings of weeds and gardens of coral. There he gave fabulous banquets for other immortals. Whenever he left his underwater domain to rise to the surface, he wore a breastplate of gold and rode in a chariot pulled by six fiery chargers with tossing manes, holding a trident in one hand as a symbol of his authority. He was accompanied by bare-breasted sea nymphs, or Nereides, supported by Tritons, who sang his praises and blew sonorous conch shells. His soldiers were sea dragons, hippocamps, centaurs, and other mythical beasts. If ancient sailors really believed that such creatures existed, it's a wonder any of them ever sailed the high seas.

The Mediterranean is known as the "cradle of civilization" because along or near its shores grew up the world's first great cities. Since prehistoric times men have put to sea from them in search of exotic products which they could not get at home. Interestingly, the oldest evidence of human navigation are wooden oars found in Denmark during excavation of the Maglemosean culture (7500 B.C.), and not in the Mediterranean. Although the early history of watercraft is speculative, scholars now believe that Stone Age man may have used inflated animal skins, wooden rafts, and dugouts for crossing rivers and lakes and possibly for making short sea voyages within sight of the coast. Archeological evidence found on the mainland of Greece reveals that more than 9,000 years ago, sailors had a type of vessel that was capable of making voyages in the open Aegean. These early mariners sailed to the island of Melos to obtain obsidian, a hard volcanic stone used to make sharp knives, scrapers, and other tools. Unfortunately, there's no record of what they sailed in. A clay model of the oldest known sailing ship, dating to about 3500 B.C., has been found in a grave at Eridu in southern Mesopotamia. Similar vessels rigged with large square sails appear in prehistoric Egyptian art, scratched or painted on pottery. The Gerzeans, who emigrated from Asia to Egypt during this period, provide us with the earliest known representation of boats, which were made from planks.

For a long time man was content to hug the coasts of the Mediterranean, believing the Atlantic Ocean to be an unnavigable sea of chaos covered by thick clouds and perpetual darkness, where giant sea serpents swallowed up whole ships and from which no man ever returned. The Phoenicians, who were the first to sail past the Pillars of Hercules into the Atlantic, finally dispelled this myth, but not before they built up a lucrative trade by spreading distressing tales, claiming, for instance, that the voyage from the Straits of Gibraltar to the British Isles took more than six months and that it was fraught with constant danger from constant wild storms, entangling seaweed, and fearsome monsters which devoured half the ships that ventured into the area. According to one chronicler, as late as 1406, four English ships bound for Bordeaux were dragged under by a whirlpool created by a giant sea monster which ate the unfortunate mariners and spat out their bones.

Men have feared the sea as long as they have sailed on it, but still, in all that time, one of their fondest dreams has been to descend to the bottom of the sea to solve its mysteries or gather its treasures. Although men have been "going down to the sea in ships" for thousands of years, they have also been going *beneath* the sea—in everything from birthday suits to barrels to bathyscaphes—for a very long time.

1

FREE DIVERS
IN THE OLD WORLD

Who the first divers were, we don't know. Nor can we be sure why they dived, though it's fairly safe to assume that fishing for mollusks, crustaceans, and other food was a major reason. We do know that men were diving as early as 4500 B.C.; archeologists have unearthed shells in Mesopotamia that must have come up from the seafloor. But there is a gap in the history of diving of more than a thousand years—until the Theban VI Dynasty in Egypt, around 3200 B.C. The vast number of carved mother-of-pearl ornaments discovered at many of the archeological sites there indicates that diving was widespread, for the shells the ornaments were made from could have been obtained only by diving.

The Cretans, who were flourishing around 2500 B.C., worshiped the god, Glaucus. Glaucus is still the patron saint of Greek divers, fishermen, and sailors today. Before he became a god, Glaucus was a fisherman in Anthedon, a village famous for its inhabitants' love of diving. It was said that the reason Glaucus became immortal was that he discovered seaweed with magical properties. Although apparently no one else had such good luck, the early Greek divers were extremely successful, for they provided the known world with most of its sponges, something their descendants did until the advent of synthetic sponges in recent years. In the ancient world, sponges were used for various purposes. They were soaked in water; made into canteens; saturated with honey and given to infants; used as padding in armor and medically to cover wounds; and, of course, women used them in the home, just as they do today.

There are many references to sponges, both in the *Iliad* and the *Odyssey*. Early Greek sponge divers apparently were the first to systematically explore the depths of the sea and bring back sponges. They also brought back man's first knowledge of the sea, some of which was distorted by contemporary writers. The divers were considered the most courageous men of their time, far more than the warriors who showed bravery in battle. "No ordeal is more terrible than that of the sponge divers and no labor is more arduous for men," said the Greek poet, Oppian, in the second century A.D., writing about the sufferings of the sponge divers of his day.

The first diver to go down attaches a cord to his middle. He also takes oil into his mouth, puts oil into his auditory canals, soaks sponges in oil and places them over his ears. He has only two tools and these he holds one in each hand: a very sharp curved knife something like a bill-hook, and a heavy stone. Taking a deep breath of the air which he is about to leave, he then dives. Thanks to the weight of the heavy stone he has with him, he plummets down rapidly. Despite the fact that his ears are full of oil and covered by oil-soaked sponges, they begin to pain him. At the same time his temples, his eyes and his chest are taken as though in a liquid vise. He hits the bottom rather than lands on it. Then he spits out a little of the oil he has in his mouth and this rises to the surface, spreading out and calming the surface agitation and lighting up the waters as a torch lights up the darkness and allows man to see in the middle of the night. He then makes for the rocks and discovers the sponges he is seeking. They grow on the underwater rocks and they seem to be part of the submerged reefs. One might think they were animated with the breath of life. Without losing a moment, he darts at the sponges, vigorously wielding his knife, which is something like a sickle. As soon as he has severed the sponge from the rock, he pulls the cord to let his companions above know that he must now be pulled to the surface as quickly as possible. Once cut away from the rock, the sponge bleeds a nauseating liquid which spreads around the diver and is sometimes sufficient to kill him, so offensive is the smell to the nostrils of man. That is the reason why the diver is eager to depart, and now his companions draw him up to the surface as rapidly as possible. The diver is now out of the water once again, but it is impossible to look at him without sympathy. The joy of seeing him once again is mingled with sorrow at observing how exhausted and at the end of his strength he seems, so much have fear, fatigue and suffering affected his vitality. Sometimes his efforts end miserably and cruelly. Once he has plunged into the waves, the unfortunate often does not return. He has encountered some hidden monster of the deep and lost his life. Desperately at first he pulls the rope in order to be hoisted to safety, but the monster seizes him and then a horrible tug of war takes place. The monster holds him from below and his friends pull him from above, disputing the half-devoured corpse of the unfortunate diver between them. Then with heavy hearts his companions hasten away from the ill-fated area, abandoning a hopeless undertaking. In tears around the remains of their dead comrade, they carry him back to shore.

The ancient art of free diving for sponges is still practiced around the Greek islands of Symi, Nisyros, and Kalynmos, the only concession to the modern age being the use of a face mask. Helmet divers came to the Aegean in the 1860s, 25 years after the surface-supplied air helmet diving suit was invented. But the free divers persisted. In 1900, of the 150 sponge boats that were based at Symi, half were still using free divers; but by 1950 only some 30 boats were operating with free divers. The prowess of the divers was mentioned by E. D. Clark, a traveler in 1837, who wrote: "In Symi, whose inhabitants are principally maintained by the occupation of diving for sponges, the following singular custom is observed. When a man of any property intends to have his daughter married, he appoints a certain day. Then all the young, unmarried men repair to the seaside, where they strip themselves in the presence of the father and his daughter and begin diving. He who goes deepest into the sea, and remains longest under the water, obtains the lady."

Besides sponges, the early divers brought up valuable shells from the sea bottom—mother-of-pearl, oysters containing pearls, and murex shells, the source of a purple dye for cloth. But the most highly prized find was red coral, which is still in demand today for jewelry and ornaments. It was largely to obtain red coral that trade was established between the Mediterranean trading nations and China. The Chinese were well aware of such treasures in the sea. As early as 2250 B.C., Emperor Yu's divers were bringing him pearls as tribute.

There is documentary proof that there was large-scale pearl diving in India and Ceylon around 550 B.C. and that vast amounts of the pearls recovered reached the Mediterranean through trade and commerce. Other than this, little is known about the early pearl divers. Passing through Ceylon in 1599, Marco Polo wrote: "The Pearl oysters are taken by men who go down in water as much as fifteen or twenty sailors arms [fathoms]. These local men, who are expert at this, are under some suspicion of being sorcerers because of the way they defend themselves from sharks, who do not touch or do any harm to those pearl fishers, whereas anyone else scarcely enters the water before they seize him."

Today there are more than 5,000 pearling vessels operating in the oyster beds of the Persian Gulf. Before they leave port, sailors and divers coat their upper hulls with evil-smelling shark oil and paint the underwater sections with lime and sheep fat to retard barnacles and ward off sharks. Each boat usually carries between 20 and 40 free divers who still use the same method of diving described by Oppian. Their only concession to the space age is to use a nose clip. Each diver makes an average of 50 dives a day, usually to a depth of up to 90 feet, where he spends two or three minutes each time. Heat, scurvy, skin diseases, and attack by sharks take a heavy toll each year. Although one Gulf pearl sold for $75,000 in 1929, diving is rarely a way to get rich. After the owners and the seamen have taken their share of the profits, the divers barely have enough left to pay their operating costs.

In Japan, the Ama divers have been at their profession since before the time of Christ. The divers are usually women, because their thicker layer of subcutaneous fat makes them less susceptible to the cold water. The approximately 7,000 Ama are slightly taller and heavier than the average Japanese woman, but they are more likely to suffer from ear disorders. Using only goggles, they duplicate the feats of their Persian Gulf counterparts. In addition to pearl oysters, they dive for shellfish and edible seaweed. Few of their daughters, who prefer to work in the cities, are following in their footsteps, and they are a rapidly vanishing breed.

In the ancient world, the work of divers didn't stop with the acquisition of valuables from the sea. Divers descended for other reasons—construction work in harbors and rivers, recovering sunken treasure, laying mines. Occasionally they were given bizarre tasks, as an amusing story of the Greek historian, Plutarch, reveals. The Roman general, Marc Antony, was persuaded by Cleopatra to partic-ipate in a fishing contest. Knowing himself to be a poor fisherman and not wanting to lose face before the Queen of Egypt, he hired a diver to keep his hook well supplied with fish. Unfortunately,

Cleopatra's spies discovered the trick. Antony's face must have been very red indeed when his first catch on the second day of the contest turned out to be a large, dead fish—salted, and ready for the frying pan.

Plutarch's anecdote suggests that there were plenty of divers around for practically any task. It's likely that the main source of employment for them, aside from bringing valuables up from the seafloor, was underwater construction in shipping and anchoring areas. So keen was the competition among the many divers engaged in this work that they formed corporations to obtain contracts for jobs. For example, during the second century B.C., the Roman emperor granted a concession to conduct all diving operations along the Tiber to one such corporation.

The earliest account of using divers to hunt for sunken treasure was told by Herodotus, a Greek historian writing in the mid-fifth century B.C. Some 50 years earlier, a Greek diver named Scyllias, and his daughter, Cyane, had been employed by Xerxes, King of Persia, to recover an immense treasure from some Persian galleys sunk during a battle with the Greek fleet. They recovered the treasure, but Xerxes refused to give them the promised reward. Instead, he kept them aboard his galley, no doubt for other diving jobs. Seething at this treachery, Scyllias and Cyane jumped overboard during a storm and cut the anchor cables of the Persian ships, which caused several collisions. As soon as order had been restored, the Persians pursued Scyllias and Cyane, but they escaped by swimming to Artemisium, about nine miles away, completely underwater.

Herodotus had some doubts about those nine miles. Such a feat was unheard of then. But later historians were impressed. They think Scyllias and Cyane may have used hollow reeds to breathe air from the surface—the forerunner of the snorkel. Such a device was first mentioned by Pliny the Elder, a Roman scholar and naturalist of the first century A.D., but it may well have been used earlier.

Diving for sunken treasure had become so common among the Greeks of the third century B.C. that special laws were passed, regulating the division of the finds. A diver who recovered treasure in two cubits (about one and a half feet) or less was entitled to a tenth of its value; treasure recovered between two and eight cubits entitled him to a third; and treasure recovered at depths of more than eight cubits entitled the diver to half of its value. The part of the treasure not given to the diver was the property of its original owner, or if the owner was dead or couldn't be found, the treasure became the property of the ruler from whose waters it was recovered.

In 169 B.C., during Rome's wars against the Greeks, Perseus, the last king of Macedonia, was defeated at the battle of Pydna. Perseus fled to Pella where he had all of his treasures sunk in the deep Lake Lydias. Then, because the Romans were no longer pursuing him, he had skilled divers raise it again. To make sure no one would be left to tell of his great panic, Perseus had the divers put to death. All of the other witnesses suffered a similar fate, but even with such ruthlessness, it proved impossible to hush up the incident.

The earliest record of divers engaging in military operations was Homer's mention in the *Iliad* of their use during the Trojan Wars (1194–84 B.C.). What their tasks were, Homer didn't say, but they probably included boring holes in the bottoms of enemy ships and cutting their anchor cables. Snorkels may have been used; concealment underwater was very useful in surprising the enemy.

The Greek historian Thucydides gives a contemporary account of military divers during the fifth century B.C. A band of Spartans on the island of Sphacteria off the southwestern coast of Greece found themselves besieged by the Athenians. Cut off from the supplies on which their survival depended, the Spartans used divers who at night swam underwater past the besiegers' ships and returned, carrying animal skins filled with food from their allies. Expecting the Spartans to be quickly starved into surrending, the Athenans were at first baffled, but they soon caught on and stationed guards to capture the divers.

Less than a decade later, the Athenians attacked the harbor at Syracuse, only to discover that the defenders had built underwater obstructions to keep out the besiegers' ships. So the Athenians sent divers equipped with saws and axes to cut down the obstructions and tie the fragments to towlines so they could be pulled out of the way. But as fast as the Athenians removed the obstructions, the Syracusan divers replaced them with new obstructions. Finally, the Athenians, originally the attackers, were themselves attacked when allies of Syracuse arrived, and the Athenians lost many ships and more than 40,000 men.

There are other accounts of divers cooperating with armies. Much of the cooperation involved sabotaging enemy ships and reconnoitering enemy ports to ascertain naval strength or eavesdrop on the conversations of enemy leaders aboard their vessels. These divers undoubtedly turned the tide of many a sea battle. By the fifth century B.C. they were considered so important militarily that the war-minded Romans were forced to take precautions against them. All anchor cables on Roman ships were made of iron chain, and special guards of the fleet, who were actually divers armed with trident spears, were on duty round the clock to prevent underwater infiltration by enemy divers.

One of the most intensive and frustrating underwater searches took place about A.D. 750. Harun al-Rashid, caliph of Baghdad and the powerful Islamic counterpart of Charlemagne, had a large number of free divers among his regular slaves. The caliph also had a ring containing a dazzling ruby. Believed to possess magical power, the ring had for centuries been handed down from one king to the next. When it was placed in a dark room the ruby would blaze like a torch, illuminating the room with its rays. The observers would notice that there were vague and mysterious figures within the stone. Upon his accession to the throne at the age of 22, Rashid ordered one of his ministers to throw the ring into the Tigris River. But soon thereafter, he decided that he had made an awful mistake and ordered that it be recovered at any cost. For five long, miserable months 500 divers worked day and night, and the ring was finally recovered. Some 100 of the divers were drowned during the search and an equal number died soon afterward from fatigue.

Then there were the Vikings, many of whom were proficient free divers. "They were able to hold their breaths under water for a long time so that during battle they could swim unseen towards the enemy with big strong drills to bore holes in the enemy hulls, thereby causing them to sink." They must have been a hardy bunch of men, as well, to swim in those cold northern waters. Using this method of undersea warfare in the year 1000 some Vikings from Sweden actually sank a fleet of 16 Danish pirate ships and killed most of the pirates.

At the end of the twelfth century an Englishman named Walter Mapes described a free diver named Nicolo Pese, or Nicholas the Fish:

His fame and skill was spread far and wide, who was able, thanks to a life spent almost constantly underwater, to penetrate the secrets of the sea so completely that he could even foretell storms. Some people not realizing what practice could do, did not believe that such a thing were possible without the aid of magic. The King of Sicily sent him with some letters to the King of Calabria and this king offered him a golden cup to explore the terrible gulf of Charybdis. There he remained three-quarters of an hour amidst the foaming abyss; on his return he described all the horrors of the place, and so astonished the monarch, that he requested him to dive once more, further to ascertain its forms and contents. He hesitated, but upon the promise of a still larger cup and a purse of gold he was tempted to plunge again into the gulf, whence he never more emerged. It is believed that he was devoured by fish when he dived into the unfathomable depths.

Another author writing about the same time claims that Nicolo survived the dive: "As an object of great interest he was taken to the court of King William of Calabria, where, being separated from that which had become his element, he gradually pined away and died." The Englishman must have grossly exaggerated the length of time he stayed down, since it would be impossible to "remain three-quarters of an hour amidst the foaming abyss."

Throughout antiquity, however, there were references to divers holding their breath underwater for long periods, a feat that still seems impossible. A Dutch visitor to Japan in 1584 wrote: "All along the coast we found women-divers, who lived with their households and families, in boats upon the water. These women dive in eight fathoms of water and more and spent the better part of 15 or 20 long minutes under water. Their eyes, by continual diving, grow red as blood; whereby a diving-woman is distinguished from all others." In 1592 another Dutchman living in Java wrote: "The greatest length of time that pearl-divers in these parts can hold under water, is about a quarter of an hour, and by no other means than custom. For pearl-diving lasteth not above six weeks, and the divers stay a while longer under water at the end of the season than at the beginning." About a century later, still another Dutchman in Java wrote: "Here at Batavia is an expert diver who draws wages for nothing else but diving for anchors, guns, etc., lost in the Roads. I have seen him several times go down, holding my breath as long as I could, but he stayed down ten times as long under water as I could hold my breath on land. But he will not go down unless you give him a whole pint of strong spirits."

For lack of evidence to the contrary, we are inclined to believe either that all the writers of the past were liars or that man's physical capacity has drastically declined over the centuries. Some Ama divers can spend four or five minutes underwater, and the record for holding one's breath underwater, set in a swimming pool with the diver completely motionless, is 13 minutes. How could anyone stay under 15 to 20 minutes?

The early divers, of course, had nothing like the equipment that the modern diver takes for granted, yet they were remarkably skillful. In the Mediterranean, red coral is seldom found in water less than 100 feet deep, and diving to such a depth without special breathing apparatus is no mean feat. A dive to 100 feet must have taken at least three or four minutes; from the many eyewitness reports, though, we know that they went deeper. Some present day Ama divers can reach 150 feet unaided, but they certainly can't do this all day long, as the free divers of long ago are said to have done. Our admiration grows when we realize that they usually performed with no protection for their eyes. Anyone who has attempted to open his eyes in salt water knows how difficult it is. On top of this, how could they have endured the cold water for so long? Even today, few divers can spend more than half an hour on the surface of the Mediterranean without a rubber diving suit.

The earliest records of free divers trying to counteract the pressure on their ears and Eustachian tubes dates from the mid-sixteenth century when a visitor to the pearl fisheries in the Persian Gulf wrote: "The divers cut out an object made from tortoise shell, something like a pair of scissors, to clip together their nostrils, prior to diving to the great depths."

There was only one other noteworthy aid to divers during the age of free diving: goggles. Who invented them and where they were first used, we don't know. A ceramic vase from second-century Peru, now on exhibit in the Museum of Natural History in New York, shows a diver wearing goggles and holding two fish in his hands. The first written reference to them is in the reports of fourteenth-century travelers returning to Europe from the Persian Gulf, where pearl divers were using goggles with lenses of ground tortoise shell. The Polynesians, whose underwater exploits are legendary, were also using goggles several centuries before European explorers discovered them. Yet free divers never used them to any great extent, probably for the same reason divers don't use them much today: while they give the diver better underwater vision, the deeper he goes, the harder the water presses against the goggles and the harder they press against the diver's eyes. This pressure can eventually result in severe and permanent damage to his eyes. Today diving goggles are used by children in shallow water and by long-distance swimmers. Serious divers prefer the face mask which covers the nose in addition to the eyes. As the water exerts pressure on the mask, a diver blows air through his nose into the mask and thus prevents it from flattening against his face.

Pearl shells recovered from the Red Sea by early free divers. They were uncovered during an archeological excavation in Egypt and date from the period of Sesostris I (circa 1971–1928 B.C.).

Mediterranean free divers harvesting sponges.

Peruvian vase dating from second or third century A.D. *showing a diver wearing goggles and holding fish.*

2

FREE DIVERS
IN THE NEW WORLD

Nearly as famous as the free divers of the Mediterranean were the pearl divers of the Caribbean. They remained free divers long after the introduction of modern diving equipment. Although this equipment made the free divers virtually obsolete, as recently as the 1940s they were enthusiastically plying the trade their ancestors had been engaged in for centuries.

Aborigines in the New World were diving well before Columbus's discovery of it, and like their Mediterranean counterparts, they too found the seafloor a good source of food. For North American Indians, diving was a basic hunting technique. Swimming underwater and breathing through reeds, they could sneak up on unwary game and capture it with nets, spears, or even their bare hands. Like the ancient Greeks, the Mayans of Mexico venerated a diving god. A fresco of their deity may be seen today in the Temple of the Diving God at the archeological site of Tulum on the eastern coast of Yucatán.

The Peruvian pot depicting a diver, which was mentioned in Chapter One, shows that diving was going on very early in the history of the New World. When the Spanish explorers first reached Tierra del Fuego on the southernmost tip of South America, they found expert women divers among the Yahgan Indians. The women would descend to great depths in the 42-degree waters to gather clams, crabs, and other seafood.

Diving for pearls didn't become a major occupation until the arrival of the white man, however, although it had been done on a small scale in the Caribbean by the Lucayan, Carib, and Arawak tribes. During Columbus' third voyage to the New World in 1498, his fleet anchored one day at the island of Cubagua, near the coast of Venezuela, to obtain fresh water and fruit. Some of his men went ashore where they noticed a Carib Indian woman wearing a necklace made of pearls. After making inquiries, they told Columbus that the natives of Cubagua possessed great quantities of valuable and exquisite pearls which were to be found in the waters all around the island. Columbus sent Indian divers in search of oysters, and the result confirmed what his men had said. When he returned to Spain, Columbus reported his find to the king who ordered the establishment of a pearl fishery on Cubagua. During the next few years, large oyster beds were found near Cubagua and especially on Margarita Island which eventually became the center of the pearl industry, a position it holds today. So Columbus, already assured of a place in history, had another discovery to his credit. Over the centuries, Caribbean pearl fisheries furnished Spain with a source of wealth surpassed only by the gold and silver that the Spaniards took from the New World.

Not long after the pearl fisheries at Cubagua and Margarita were opened, the supply of local divers was exhausted. Many died from diseases brought from Europe by the Spaniards, while others died from exhaustion brought on by being forced to dive as many as 16 hours a day. The Spaniards then turned to the Lucayan Indians in the Bahamas as a source of divers. The Lucayans were considered the best divers in the New World. The Spanish historian, Oviedo, writing in 1535, wrote an account of a visit to the pearl fisheries at Margarita where he watched the divers. He marveled at their abilities, saying they could descend to depths of nearly a hundred feet, remain submerged as long as 15 minutes, and, unlike the Carib Indians who had less stamina, dive from sunrise to sunset seven days a week without appearing to tire. As the divers of the Old World had been doing for thousands of years, these divers descended by holding stone weights in their arms and jumping overboard, naked except for a net bag around their necks, in which they would deposit the oysters they found on the seafloor. So much in demand were the Lucayan Indians that within a few years nearly all of them were enslaved, and before much longer, the first natives Columbus saw on his epic voyage of discovery had vanished.

By 1550 the Spaniards were again hard pressed for divers. To solve the problem, they imported Negroes from Africa, many of whom had never seen the ocean, much less, dived. Surprisingly, they adapted almost immediately, soon becoming as good at diving as their predecessors had been. Women were preferred to men—probably, like the Ama, because of the extra body fat—yet, regardless of sex, their average working life was only a few years. Like the Indians before them, they suffered from overwork and disease. As though this weren't bad enough, cannibalistic Carib Indians regularly descended on the pearl fisheries and carried off large numbers of them. They were also subject to attack by sharks and other denizens of the deep. Writing in 1618, the governor of Margarita Island said: "So few divers dare venture out these days without being threatened with instant punishment, for just this past year more than two dozen were devoured by sharks and during the twenty odd years that I have resided on this island no fewer than 200 have met a like fate." And a visitor to the island in 1693 wrote:

The divers here, who number some 500 or more, are among the best in the world for these dive to depths of 15 fathoms and spend as long as 15 minutes beneath the sea. They are exposed to a great deal of danger from the large fishes, which are numerous here, called tiburones [sharks]; they are of a monstrous size, very fierce and voracious and they often devour the poor divers. Another sort called manta rays, which are the size of a large blanket, are like monsterous thick thornbacks and they embrace the diver so strongly that they squeeze them to death, or else by falling on them with their whole weight they crush them to death against the bottom. In some measure to secure themselves from these fishes, each diver takes along with him a sharp knife, [with] which he wounds the fishes and puts them to flight. The Negro overseer who stays in the boat watches constantly through the clear waters and when he sees any of these fishes making towards them, gives them notice by pulling on the line which is fastened to the diver and at the same time, takes some weapon and dives to their assistance. But notwithstanding all these precautions, the Negro divers sometimes lose their lives and often a leg or arm. The pearls found here are among the most cherished in the whole world and that is the reason the kings of Spain feel no sorrow about forcing these unfortunate Negros to risk their miserable lives in pursuit of the pearl oysters.

The most interesting thing about the Caribbean divers was their ability to stay underwater so long. While Oviedo's mention of 15 minutes appears to be an exaggeration, there are at least six later accounts by travelers in the Caribbean who witnessed the Margarita pearl divers in action, and all say the same thing. Were they merely echoing Oviedo or did the Caribbean divers really possess a long-lost secret which enabled them to stay underwater that long?

The divers believed they owed their remarkable endurance to, of all things, tobacco. They were very heavy smokers. In 1617 the governor of Margarita wrote to the king of Spain, saying that when the island had run out of tobacco, the divers went on strike. The governor first tried punishing the divers to get them to go back to work but finally gave up and sent a ship to Cuba for a new supply of divers.

Whatever the Caribbean divers of three centuries believed about the virtues of tobacco, today's would-be diver is not advised to take up smoking to build up his endurance. Medical science has proved conclusively that it has just the opposite effect.

While it's difficult to believe the tobacco legend, there is another one concerning the divers of the Caribbean, which is less of a strain on our credulity. In 1712 the current governor of Margarita wrote that during his 50 years on the island, many divers had been killed by sharks "and other monsters," but that a few years earlier, the divers had found a mineral which, when rubbed over their bodies before they dived, would repel sharks. Unfortunately for us today, who are still trying to find satisfactory shark repellent, the governor didn't say what the mineral was, so there is some doubt about this legend, too.

In 1738 another European visitor to the Margarita pearl fisheries wrote an even more pathetic account of the dangers and hardships the divers faced:

> This business of a diver, which appears so extraordinary and full of danger to a European, becomes quite familiar to a Negro, owing to the natural suppleness of his limbs, and his habits from infancy. His chief terror and risk arrive from falling prey to the large sharks which frequent these waters. Only young Negros who can hold their breath for a long time are suitable for the work. They fill their mouths with coconut oil which they spit out into the water and this gives them a moment of breath. Upon reaching depths of 20 fathoms, where more of the oysters

are found, the diver spends around 15 minutes filling the net hung around his neck with the oysters. The exertion undergone during this process is so violent, that upon being brought into the boat, the divers discharge water from their mouths, ears and nostrils, and frequently even blood. But this does not hinder them from going down again in their turn. They will often make forty or fifty plunges in one day, and at each plunge bring up about a hundred oysters. Quite often these divers will suddenly drop dead from haemorraging or congestion. It is work that cannot be done for more than four or five years in succession. Having reached the age of twenty-four they cannot hold their breaths long enough anymore. A good diver does not eat much and always only dried food and never any type of seafood for they feel that it can cause them to drown.

Shortly after Balboa discovered the Pacific Ocean by crossing the Isthmus of Panama, Spanish explorers found fabulous pearl beds around the Pearl Islands south of Panama City. In an attempt to appease the white invaders, the Indians gave them hundreds of pounds of pearls, "some of them as big as hazel-nuts." By 1540 the Spaniards had established a pearl fishery in this area which rivaled the one off the coast of Margarita. Within 50 years all of the Indian divers were used up and Negroes from Africa were brought in to replace them. A visitor in 1750 wrote: "Those only are accounted compleat divers who have kept themselves under water till the blood gushes from their eyes, mouth and nose. This accident is said never to happen a second time and is not dangerous, the hemorrhage stopping itself. The divers have more fear from the sharks, manta rays, and poisonous sea snakes—this latter type of sea fish accounts for more deaths than the other two. Generally after one year of diving the divers lose complete facilities of hearing and their eye sight is poor except when beneath the sea."

There is little in the stories about the Caribbean divers for us to take seriously today, but their stamina must command our respect. The Spaniards, who knew when they were onto a good thing, soon found a use for talents as important as pearl diving: salvage work. From the time the Spanish Crown opened the New World to others besides Columbus in 1503, many ships sailed each year from Spain carrying supplies to settlements in the New World. On the return voyages they carried the treasures and products of the New World back to Spain. But, due to poor seamanship, faulty navigation, storms, or all of these, many of them were lost at sea. In such major colonial ports as Havana, Veracruz, Cartagena, and Panama, teams of native divers were kept aboard salvage vessels which were ready to depart on short notice to attempt the recovery of sunken treasure. From the sixteenth century to the eighteenth, more than 100 million ducats were recovered from Spanish wrecks by these divers who, on more than one occasion, saved Spain from bankruptcy. Ironically, when the other European nations began colonizing the West Indies, the same divers were instrumental in depleting the Spanish treasury: their new employers used them to salvage Spanish wrecks. But this time, the profits went into the English, French, and Dutch treasuries.

By the middle of the sixteenth century, when the Spaniards were forced to send their ships across the Atlantic in organized convoys, called *flotas,* because of the increasing number of attacks by pirates and enemy fleets, every ship in a *flota* carried West Indian or Negro divers who proved invaluable at every stage of the voyage. Before a *flota* left port, divers would inspect the ships and make necessary underwater repairs. So highly regarded were they that an adverse report on a ship's condition was enough to prevent the ship's sailing. There is no doubt that they were responsible for saving many ships and their cargoes. Once a *flota* was underway, the divers were in constant demand. Since the ships routinely sailed dangerously overloaded and in all kinds of weather, not infrequently their seams would open, causing them to leak badly. Once a leak had been located, the divers were lowered over the side where they would seal the leak with wedges of lead or nail large planks over it. Neither method was easy. A ship couldn't stop or it would fall behind and lose the protection of the convoy. An admiral of a *flota* in 1578 wrote to the King, recommending the conferring of a title on one of his divers who, by keeping several ships from sinking, deserved much of the credit for the safe arrival of 12 million ducats in Spain that year.

Another task assigned to divers—and a profitable one—was a kind of underwater customs duty. When the treasure-laden *flotas* reached Spain from the New World, a great deal of smuggling ensued. To avoid the high taxes, smugglers came up with various ways to outwit the King's customs officials. One of the most common tactics was for dishonest divers to attach part of the cargo to the underside of a ship. Another was to throw these hoards overboard before the customs officials arrived and recover them later by diving. Rarely did a year pass in which divers working for the King did not find at least one large treasure hoard. On one occasion a diver discovered that an enterprising captain had had the lower part of his ship's rudder made of solid silver while in port in the New World. The fraud was detected when the diver noticed that the paint concealing the silver had worn off during the voyage. Presumably, he was well rewarded. Divers engaged in this work were usually well paid; sometimes they earned enough to buy their freedom.

Settling on Bermuda in 1609, the English soon realized the importance of the Negro pearl divers. Shortly after Bermuda was first settled, privateers from Bermuda were raiding the pearl fisheries where they captured a large number of divers and put them to work salvaging wrecked Spanish treasure ships. These divers fared much better; many were given their freedom in return for salvaging valuable sunken treasure. Until the mid-seventeenth century, when Port Royal, on the island of Jamaica, was founded, treasure-hunting, or "wracking," as the English called it, was the major industry on Bermuda. The settlers had dozens of sloops and schooners working wrecks throughout the Caribbean.

Due to its strategic location near the middle of the Caribbean, Port Royal quickly became the headquarters of the "wrackers." A Spanish spy who sneaked into Port Royal in 1673 reported that as many as 50 sloops and schooners were operating out of the port, hunting for treasure from Spanish shipwrecks. The lure of sunken treasure has long drawn men to the depths of the sea. As might be expected, Spain, which lost countless ships during the sixteenth and seventeenth centuries, took the lead in recovering treasure. Yet despite this Spanish effort, the greatest recovery of treasure from a single vessel until the present century was made by an American, and to this day, the old veteran divers of Port Royal are honored for having recovered the largest amount of treasure from a single shipwreck.

To add to the fairy-tale quality of the story, the ship involved was Spanish, a member of the *flota* of 1641, one of the richest ever to sail from the New World for Spain. The *flota* consisted of eight heavily laden galleons. But six of them developed leaks even before sailing, and most of the gold and silver bullion (valued at more than 20 million ducats) had to be carried by only two galleons—the flagship, *Capitana*, and the vice-admiral's ship, the *Almiranta*. Traveling with the treasure galleons under convoy were 22 merchant ships loaded with various products from the New World, including tobacco, chocolate, lumber, and sugar.

Two days out of Havana, while passing through the Florida Straits, the *flota* was struck by a severe hurricane. Within a few hours all of the ships had sunk, except for the *Capitana* which later sank off the coast of Spain and the *Almiranta* which was filled with water and barely stayed afloat, losing its mast, sails, and rigging. In a frantic effort to keep the *Almiranta* from sinking, the officers, crew, and passengers bailed around the clock during the hurricane and its aftermath. For nearly a week the ship was carried along by the wind and currents, and at last, some 50 nautical miles north of Hispaniola, struck a reef. Most of the 600 passengers managed to swim to a nearby sand spit where their chances of survival seemed slim. Much of the *Almiranta's* food and water had been thrown overboard in the struggle to keep her afloat. Makeshift rafts and boats were built from the wreckage that remained above water, and in them 200 survivors sailed for Santo Domingo. Only a few reached the island.

When the Spanish government learned of the fate of the *Almiranta*, they prepared an expedition to help the survivors and try to salvage the wreck, but bad weather delayed the expedition's departure. By the time it reached the sand spit where those who didn't sail to Santo Domingo had stayed, not a living soul was left alive. All 400 or so had died of thirst, hunger, and exposure. The expedition went on to Margarita to pick up 50 pearl divers for the salvage operation, but one storm after another delayed them. It wasn't until the following spring when it was found that the numerous storms had not only completely submerged the wreck but had washed away the sand spit, so that the salvors now had no idea where it was. During the next 20 years, the Spanish mounted over 60 unsuccessful expeditions to try to find and salvage the wreck, finally admitting defeat. Thus the richest known wreck in history was lost forever—or so the Spaniards thought.

Although the rest of the world may have been content to leave the treasure on the reef, which had become known as Silver Shoals, an American named William Phips wasn't. As a child in Boston, Phips had listened to sailors' tales and was bitten by the treasure bug. In 1681 he took the money he had saved working as a shipwright and later in his own shipping business and went off on his first treasure hunt to the Caribbean. He didn't find the vast treasure of his dreams, but he did locate several wrecks in the Bahamas, which more than covered his expenses. It was a promising beginning. Certain that he would hit the jackpot soon, Phips decided to go after a Spanish galleon that reputedly carried gold and which had sunk near Nassau. For the venture, Phips wanted the best possible ship, men, and equipment, and for this he needed more money than he had. Failing to raise money in Boston, he went to London in the spring of 1682, where he hoped to get help from King Charles II. It was 18 months before Phips was granted an audience, but he was a determined man and waited it out. Eventually he persuaded Charles to back the expedition in return for a large share of the booty.

Phips spent several weeks locating the wreck which turned out to have no treasure, possibly because it had never carried any or someone had gotten there first. Undaunted, Phips decided to look for another wreck; but his crew had other ideas. Angry at not having found the treasure and thus gotten their share, they mutinied, intending to take over the ship and get rich as pirates. With only eight of the more than 100 men remaining loyal, Phips somehow managed to put down the mutiny and bring the ship into Port Royal where the mutineers were thrown in prison and a new crew signed on. While there, he heard about the *Almiranta*, lost on Silver Shoals. This was enough to send him off again. Before Phips reached the vicinity of the wreck, however, his new crew mutinied. Once again, he put down a rebellion. But this time, hoping to find a more trustworthy crew, he returned to England.

During his absence Charles II died and the new king, James II, had no interest in financing treasure hunts. James had Phips's frigate repossessed and, when Phips protested, imprisoned him for several months. Captivity did little to crush Phips's spirit or his salesmanship, however. After his release Phips persuaded the Duke of Albemarle and Sir John Marlborough to back the attempt to salvage the *Almiranta*, who, in turn, persuaded the King to join the venture. The Duke and Sir John argued so effectively that, not only did James furnish a ship, but he granted Phips's salvage company the exclusive concession for treasure hunting in the Caribbean, which thereby offended the Spanish ambassador.

Phips sailed for Silver Shoals, detouring by Jamaica where he took on about 24 Negro pearl divers, refugees from the pearl fisheries at Margarita. For several months he drove himself and his men nearly to the breaking point, particularly the divers, who worked from sunrise to sunset. But his perseverance was rewarded when in 1685 they found the wreck. As one of the divers rose to the surface with his hands full of silver coins, Phips burst into tears of joy. For a month they struggled to bring up treasure while fighting off the pirates who had heard about the find. Some 32 tons of silver, a vast amount of gold, chests of pearls, and leather bags containing other precious gems were recovered before bad weather and running out of provisions forced Phips and his men to suspend salvage operations. The value of the treasure was put at $3 million in today's currency. Phips received a sixth, which was enough to make him one of the richest men in the New World. Each Port Royal diver received a large bonus, and those not freemen were able to buy their freedom. Some of the now-rich divers decided to retire from the sea and invested their bonuses in taverns and other businesses (all of which were lost in the earthquake of 1692).

Phips was knighted and after declining a post in the British Admiralty, was named governor of the colony of Massachusetts. But Sir William Phips was living proof of a saying that was already centuries old: once the treasure bug has bitten a man, it never turns him loose. For a few years Phips performed his gubernatorial duties adequately. Then he abandoned his post to go on another treasure hunt. This time he had enough money to finance the expedition properly, but he didn't get the chance to put it to use. While in London in 1694, waiting for his ship to sail, Sir William died.

At the time of the earthquake on June 7, 1692 many Port Royal divers were in the pay of the Spanish. The year before, four Spanish galleons commanded by the Marquis de Bao, sailing from Cartagena to Havana, were, due to faulty navigation, wrecked on Pedro Shoals, 130 nautical miles south of Kingston. Several Port Royal boats in the area at the time helped rescue 776 persons from the wrecks, as well as part of the treasure and other cargo. No sooner had the survivors reached Port Royal than dozens of ships hastily sailed for the site of the wrecks. But this time, although the divers from Port Royal recovered much from the wrecks, they didn't benefit much from the venture. Spain and England were at peace, and the high-ranking Spanish officials among the survivors convinced the governor of Jamaica that it would cause ill feeling between the two countries if the divers were permitted to keep the treasure they were so feverishly salvaging. The governor was persuaded, because, as ship after ship put in at Port Royal, loaded with treasure, Admiralty officials seized everything aboard the vessels, and the salvors ended up with only a tenth of what they had salvaged. Within a month of the disaster the Spaniards had several warships stationed at the wreck site to prevent further diving. The Spaniards finally hired a lot of divers from Port Royal and, besides a reasonable salary, gave them a bonus of one-fifth of all that they recovered.

From contemporary documents, however, we know that not all the Port Royal divers were at Pedro Shoals diving for the Spaniards when the earthquake occured. Various accounts state that almost immediately after the earthquake, divers were searching the submerged buildings and recovering items of value. Some used diving bells which allowed them to remain submerged for an hour or more. The others had no diving equipment. These free divers were a fearless breed of men. It was reported that they often fought with sharks around Port Royal just for the sport, and they undoubtedly recovered much of value from the submerged buildings.

Although the section of the city between forts James and Carlisle sank to depths of 30 to 50 feet within two minutes of the third tremor, the section between Fort James and Fort Carlisle gradually sank, with the upper parts of the buildings remaining above water for years. In July 1693 a visitor wrote: "The Principal parts of Port Royal now lie four, six, or eight fathoms underwater. . . . Indeed, 'tis enough to raise melancholy thoughts in a man to see chimneys and the tops of some houses, and masts of ships and sloops, which partaked of the same fate, appear above the water, now habitations for fish."

Since most of Port Royal's roofs were made of wood and slate shingles which could easily be torn off, access to the buildings must have been a simple task, not only for the divers but for salvors who didn't have to descend into the water. The latter used two methods, dredging and fishing. To dredge, they would lower heavily weighted fishing nets and drag them across the bottom, snagging anything loose. To fish, they would first spread oil on the surface of the water to calm it so they could see to the bottom better. Then they would use either long poles with hooks or spears at the end, or grappling hooks attached to ropes.

The divers, dredgers, and fishers soon on the scene would have overlooked little of Port Royal's treasure. From the records, we know that extensive salvaging continued for several decades after the earthquake. How much treasure was lost and how much was salvaged, we can only guess at, since no records have so far been found. From my own excavation at the site, however, I have learned several important facts about the early salvors. First, they were good at their work, missing little of value. Inside three houses we've excavated, there was nothing except a grappling hook in one house and a two-pronged harpoon in another, both probably lost by the original salvors. In between standing or falling walls, rarely did we find anything valuable except such items as bottles, ceramic sherds, and clay smoking pipes, which the early salvors wouldn't have considered of any value. Luckily, those salvors didn't have the proper equipment or they considered it not worth the effort to remove the hundreds of fallen walls from old buildings. Otherwise, there would be little left today. During our excavation nearly every item of value—pewter, silver, gold, brass, and so forth—was recovered from under fallen walls. There was only one major exception: the first hoard of Spanish silver coins, found in a wooden chest with a brass keyhole plate and bearing the coat of arms of the King of Spain. The only explanation of why the chest was overlooked by the first salvors is the possibility that its great weight caused it to sink immediately deep into the mud on the seafloor where the salvors couldn't see it.

About the time the Port Royal divers had salvaged all they could from the sunken city, another golden opportunity arose. In 1715 a Spanish treasure fleet of 10 ships was wrecked on the coast of Florida between Cape Canaveral and Fort Pierce. More than a thousand people and 14 million pesos worth of treasure were lost. As soon as news of the disaster reached Havana, salvage teams were rushed to the site. Just as quickly, the news reached Port Royal, and dozens of small vessels sailed for Florida where, within sight of the Spanish salvage teams, divers began descending illegally to some of the wrecks. Contemporary accounts are contradictory and therefore confusing. Still, it's estimated that the divers from Port Royal recovered about 500 thousand pesos before the Spaniards sent a squadron of warships to prevent further diving on wrecks the Spanish considered their property alone.

Salvage operations were finally halted in 1719 even though the Spaniards had recovered only about half of the estimated treasure aboard the ships. The remainder, the Spanish recorded, had probably been covered over by shifting sands, making it impossible to find. Once again, the Port Royal salvors were presented with a fine opportunity. Each year between 1720 and 1728, ships from Port Royal searched for treasure at the wreck site; but the lack of historical records prevents us from learning how much they eventually salvaged.

Pearl diving, treasure hunting, repair and detective work—there seems no limit to the kinds of jobs entrusted to Caribbean divers in the past. For instance, in the memoirs of the French missionary Père Labat, there is a description of a diver performing an unusual task, one which must rank among the great adventure tales for sheer excitement:

> While visiting the Island of Saint Kitts, I learned from some people, whom I trust to tell the truth, that in 1676 a large hammerhead shark bit off the leg of a young boy who was swimming in the harbor and this resulted in the death of the boy. A Caribe Indian, who was a local diver very skilled in spearing fish underwater, volunteered to kill the shark. To understand the danger of this undertaking, one must first realize that the hammerhead shark is one of the most voracious, powerful, and dangerous fish in the sea. The father of the child who had been killed by the shark was glad of the opportunity of having the monster killed and thus offered the diver a good reward to obtain this poor consolation.
>
> The diver armed himself with two good well-sharpened bayonets and, after raising his courage by drinking two glasses of rum, he dived into the sea. The shark, which had acquired a taste for human flesh, attacked the diver as soon as he saw him. The diver allowed it to approach without doing anything until the moment he thought it was on the verge of making its rush. But at the instant it charged he dived underneath it and stabbed it in the belly with both bayonets. The result of this was at once made apparent by the blood which tinged the sea all red around the shark. Each time the shark rushed the diver, he repeated this same tactic and repeatedly stabbed the shark. This scene was enacted seven or eight times, and then at the end of half an hour the shark turned belly up and died.
>
> After the diver had come ashore, some people went out in a canoe and tied a rope to the shark's tail, and then the shark was towed to the beach. It proved to be 20 feet long and its girth was as large as a horse. The child's leg was found whole in its stomach.

3

DIVING BELLS

When Alexander the Great set out to conquer the world in 332 B.C., the island stronghold of Tyre resisted so long and fiercely that both attacker and defender had to use divers. On one occasion the defenders had them destroy a dike of timber which had been erected as a blockade by the Macedonians. When the Tyrians' will to resist had been reduced sufficiently, Alexander gave his divers the job of destroying the port's boom defenses. To students of diving history, the most interesting aspect of the battle is a legend that Alexander himself descended in some sort of container to watch the attack on the boom defenses. A thirteenth-century French manuscript contains an illustration showing Alexander inside a glass barrel brilliantly lit by two candles and surrounded by numerous species of marine life, with a large whale dominating the scene; but this is an imaginary reconstruction, as were all later illustrations. We just don't know what this container looked like.

Over the centuries many writers and historians have taken the story of Alexander in a "diving bell" and adapted it to their own uses. An Ethiopian version, probably inspired by Christian mysticism, portrays Alexander as a devout Christian, although he lived three centures before Christ. The most interesting version was written by an Arab historian in the seventh century A.D. Here, Alexander's divers are building the boom defenses while the enemy tries to destroy them:

> . . . one night when he was alone and thinking over these things an idea came to his mind. The following morning he called his workmen to-

Alexander the Great descending in a diving bell at Tyre (Indian miniature, ca. 1595).

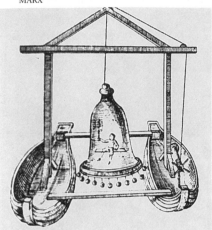

LEFT: *Metal diving bell used in Spain during the late sixteenth century.*
RIGHT: *Kessler's diving bell, 1616.*

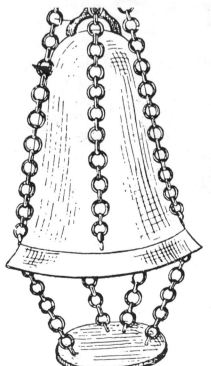
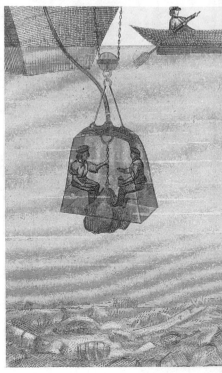

LEFT: *Seventeenth-century Spanish diving bell.*
RIGHT: *Nineteenth-century diving bell being raised along with salvaged material.*

gether and caused them to build a wooden case ten cubits long by five wide. Windows of glass were set in and the wood itself was treated with resin, wax and other substances to keep out the water of the sea. Alexander and two of his secretaries, skilled draughtsmen, then entered the case, after which he gave orders that it was to be closed up and the lid impregnated with the same substances to make it watertight. Two large ships put out to sea. Weights of iron, lead and stone had been fastened inside of the box to bear it down to the depths. Poles of wood were placed from one ship to the other and the case was then suspended on this gantry. After which the cables were paid out and the case descended into the water. Thanks to the transparency of the glass and the limpidity of the water, Alexander and his two companions were able to see the marine monsters and a species of demon having the head of a ferocious beast attached to a human body. Some of them carried axes, others saws, and still others hammers so that they looked like workmen. Alexander and his two secretaries drew careful pictures of these monsters. Then they pulled the line and at this signal the men on the ships drew up the case. The king stepped out and was carried back to Alexandria.

Scanty as the details of this exploit are, Alexander's descent is important historically, because it represents a milestone in man's struggle to exploit the sea. But it wasn't the first such technical achievement. In his *Problematum,* Aristotle mentions the use of a similar device in 360 B.C.: ". . . in order that these fishers of sponges may be supplied with a facility of respiration, a kettle is let down to them, not filled with water, but with air, which constantly assists the submerged man; it is forcibly kept upright in its descent, in order that it may be sent down at an equal level all around, to prevent the air from escaping and the water from entering. . . ." How long such devices had existed before Aristotle described them, we don't know, nor do we know how widely they may have been used. Diving devices do not appear again until the year 1250 when Roger Bacon, in his *Novum Organum,* mentions Alexander's container. Diving is thought by most authorities to have been invented in the sixteenth century.

The first known appearance of the diving bell after Alexander the Great's alleged use of it took place in 1531 when a bell was used in the lake of Nemi, near Rome, to locate two of Emperor Caligula's pleasure galleys which were said to have carried treasure. The invention of an Italian physicist named Guglielmo de Lorena, the barrel-shaped bell covered the diver's head and torso and was raised and lowered by means of ropes. The diver could walk the lake bed, remaining submerged for nearly an hour before he had to come up for air. Within a few weeks he found both wrecks which free divers had been searching for for years. Finding the galleys was only part of the problem; they still had to be raised. Although many unsuccessful attempts were made during the next three centuries, it wasn't until shortly before World War I that anyone succeeded. By draining the lake, the Italians got them raised. But the repeated failures after that of de Lorena's bell detracted little from his achievement, for his invention had done what it was designed to do, and the word spread.

In 1538 two Greeks designed and built a diving bell and demonstrated it in Toledo, Spain before the Emperor Charles V and some 10,000 spectators. Larger than de Lorena's bell, this one was large enough to hold both inventors who sat inside on planks. They carried a lighted candle inside the bell with them. To the astonishment of the king and spectators, the candle was still burning when they returned to the surface. The candle was good showmanship, but it also provided underwater illumination and—though the inventors probably didn't know this—it served as a safety gauge. If the flame went out while the divers were below, it would mean that their oxygen supply had been exhausted. The same danger that threatened Leonardo da Vinci's "SCUBA"—carbon dioxide—threatened divers who used the bell.

The news of the Toledo diving bell spread like wildfire throughout Europe, and many similar bells were built. Practical use of the bell sometimes differed from the demonstration in one important respect. When they discovered that remaining inside the bell as the Greeks had done kept them from doing the work they had been hired to do, the divers used it as a kind of air bank, swimming outside to work and returning to it from time to time for air. However the bells were used, they seemed to serve the purpose. On the other hand, some of them, such as the two inventions of an Italian named Niccolo Tartaglia, apparently served no purpose. Designed in 1551, though never built, one of them consisted of a wooden frame shaped like an hourglass on which the diver stood. A glass bowl enclosed his head, thus assuring him of good vision but no breathing to speak of, because it provided the diver with no more than a few minutes of air. The other invention featured the same hourglass frame but differed from the first, in that the diver's entire body was enclosed in the bowl, which seemed to have no opening (how the diver was supposed to c in and out wasn't explained).

The first appearance of the diving bell in the New World was in 1612 when it was used by an Englishman, Richard Norwood. Deciding that there wasn't enough money in his regular line of work —piracy—to suit him, Norwood went in search of treasure-laden wrecks which, he had heard, lay in the vicinity of Bermuda. He used a diving bell made by inverting a wine barrel and attaching weights to carry him to the bottom. Norwood didn't find the Bermuda treasure, so he went to look for wrecks in the West Indies. We don't know what success, if any, he had; but it's doubtful. A man who couldn't make a go of it as a pirate in an age when piracy flourished would not have been likely to succeed in a newer, more challenging line of work.

Meanwhile, European inventors were trying to improve the diving bells already in existence. A German named Franz Kessler perfected one in 1616 that wasn't attached by ropes to anything on the surface. The diver descended with the aid of weights which were released when he was ready to ascend. The diving chamber consisted of a wood barrel long enough to reach the diver's ankles. It was covered with leather to make it watertight and had two windows so the diver could see out. Although tests showed that Kessler's invention would work, it was clearly ahead of its time and was not widely used. In Kessler's day, people were too afraid of the ocean depths to regard his invention as anything more than the work of a crank. It was to be many years before man's dream of moving about underwater, unencumbered by lines, became an everyday occurrence.

In 1621 a famous English scientist described the operation of a diving bell used to salvage several of the Spanish galleons lost along the coasts of Scotland and Ireland in 1588:

> This vessel is made of metal, hollow like a case, and being let down with its bottom parallel to the surface of the water, it carried along with it all the air it contains to the bottom of the sea; and having three feet to stand upon, somewhat short of the height of a man, the diver, when he wants to breathe, conveys his head into the cavity of the vessel, where being refreshed with air, he afterwards continues his work. . . .

And in 1677 a wooden bell 13 feet high and 9 feet from rim to rim was built in Spain and used to salvage two rich shipwrecks at the port of Cadaqués. The size of the bell provided the two Moorish divers with enough air to remain submerged for over an hour at a time. They said they could have stayed below longer, but the terrific heat created in the bell by their breathing forced them to surface and let the heat escape. When they wanted to come up, they tugged at a line attached to the rim of the bell, and their assistants on the surface pulled them up. The divers did all their work inside the bell which was lowered directly to the wrecks. The venture was a huge success; several million Spanish pieces of eight were recovered. The divers were paid in an unusual way: each time they surfaced with chests or bags of money, they were allowed to keep as much as they could hold in their mouths and hands.

The diving bell was improved significantly in 1689 when a French physicist, Dr. Denis Papin, devised a way to supply it with fresh air from the surface by means of a large bellows, or pump. Papin's invention had four advantages. First, it permitted divers to remain below for unlimited periods of time. Second, it eliminated the danger of succumbing to carbon dioxide in the air supply. Third, because the fresh air forced out the heat, working conditions inside the bell were improved. Finally, and most important, the new design enabled the divers to reach greater depths, since pumping air into the bell kept the water out. To appreciate the importance of this last advantage, we should understand that a bell lowered with air (at atmospheric pressure) that was trapped while the bell was on the surface, was limited as to the depth it could reach. As the bell descended, the increasing water pressure would compress the air inside the bell, and water would replace it. For example, at 33 feet, water pressure equal to two atmospheres—or twice that at the surface—causes the air within a diving bell to be compressed to half the volume it is on the surface, causing the water to rise halfway up the bell. At 66 feet, or three atmospheres, the water rises three-fourths of the way up. This was the maximum depth the bell could reach with the diver's head above water. In theory, Dr. Papin's diving bell could descend indefinitely, but the pressure that pumps of his day could exert was limited. They were powerful enough to allow a bell to descend 70 feet—not much deeper than earlier diving had been able to go. For another century, until more powerful pumps were developed, Papin's invention was workable only on paper. There's no denying the soundness of his idea, though. At the very least, it paved the way for diving bells of the future.

About the time Dr. Papin was working on his diving bell, but before he had tested it, Edmund Halley, the astronomer of Halley's comet fame, built a diving bell provided with a continuous supply of fresh air, using a different method. A valve with a tube attached was installed in a lead cask. Then the cask, containing air, was lowered into the water. The divers inside could pull in the end of the tube, open the valve, and get as much air as they needed. Although more primitive than Papin's diving bell, Halley's worked better, simply because it avoided the use of those even more primitive pumps of the time.

Halley's diving bell had other advantages. For one thing, it was larger than Papin's, containing about 60 cubic feet of usable space. For another, it was made of wood and had a lead covering which kept it watertight and prevented it from overturning due to the unevenly distributed weights. There were glass viewing ports in the sides and top and an exhaust system on the top to release the diver's warm breath. Halley also provided a way to extend the radius of maneuverability underwater. The diver wore a full diving suit and a helmet to which was attached a flexible tube. The other end of the tube was held inside the bell by another diver. Initially the helmet was made of leather, but Halley replaced this with one made completely of glass, similar to the bowl of Niccolo Tartaglia's bell.

As long as there was enough air in the casks to keep them from filling with water, Halley's diving bell could descend to depths that were beyond the capabilities of others of its day. Until 1788 the casks were the principal means of providing fresh air for a diving bell. In that year an English engineer named John Smeaton built a practical, reliable pump and thus made Dr. Papin's diving bell a reality. As more and more pumps were built, Papin's method of supplying air to diving bells was widely adopted, and within a few years it had made Halley's bell obsolete.

Halley's diving bell, coupled with the recovery of the bulk of the treasure from the *Almiranta* by Phips in 1685, resulted in the springing up of numerous salvage companies in Europe. The race for quick riches was on. But the only truly effective salvors were the Spanish who usually recovered treasure from their own ships after they had gone down at sea, but while they were still visible above water. The fact that others had only limited success was attested to by the English scientist, Sir Hans Sloane, in 1707:

> I remember an African ship, wrecked on the coast of Sussex, loaded with elephant teeth [ivory tusks], which Mr. Halley told me was in a very short time almost covered with sand and mud, so that the project of recovering the teeth was frustrated, though by the help of the diving bell, contrived by his extraordinary skill, they had gone to the bottom of the sea and salvaged all aboard the hulk. Though the money brought into England from the Spanish wreck salvaged by Phips was very considerable, yet much more was lost on projects of the same nature. For every silly story of a rich ship lost, a patent taken out, divers, who are used to pearl fishing and can stay underwater some minutes, bought or hired at great rates, and a ship set out for bringing home the silver. There was one ship lost amongst the reefs of Bermuda which was very rich. It is said to be in the possession of the devil and I have heard many stories how he kept it. I do not find the people, who spend their money on this, or any of these projects, excepting the first [Phips], got anything by them.

The London *Annual Register* of August 29, 1775 gives another account of the use of Halley's bell in an unsuccessful treasure hunt:

> By letters from Rome of this day, they had ended their third trial of searching in the Tiber for antiquities and with the same bad success of not having a halfpenny profit, though they had this year an English chain pump, that did for its part wonders, in throwing out the water, but it seems that all the pumps in the navy would not answer the purpose, as the water leaked in as fast as it was thrown out. Thus, if they make any more trials, it must be in the manner they should have begun by, that is by scooping up the dirt, as done in rivers and harbours, to keep them clear; but it was presumed that they would want a new subscription for it and few would contribute after so many unsuccessful trials. We cannot, however, forbear recommending the trial of Dr. Halley's diving bell on the occasion. The leakage, which has hitherto proved so fatal, is in all probability from the bottom. Now Dr. Halley's diving bell may be cleared of water within a very small way of its lower rim, and this lower rim brought so close to the bottom, if any way even, as to afford the workmen the same opportunity of digging, which they would have in a piece of ground overflowed with water to a small depth. Nay, the bell might be lowered, with the same advantages, in pursuit of treasure, into the hole itself, let it be ever so deep, if made large enough to the purpose.

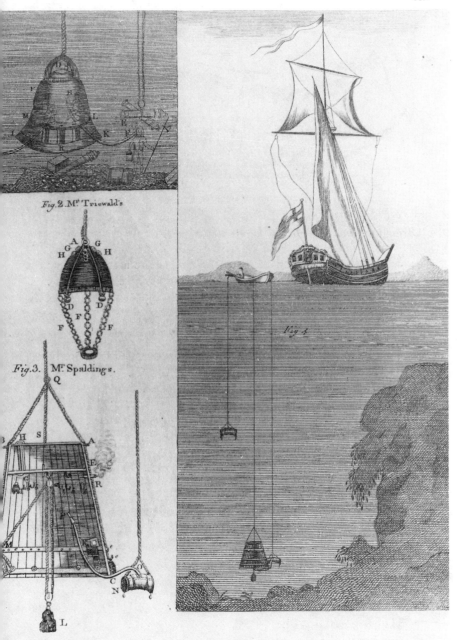

Diving bells, a copper plate engraving published for the New Encyclopedia, *by I. Low, N.Y., 1810. Shows bells developed by Halley, Triewald and
Spaulding.*

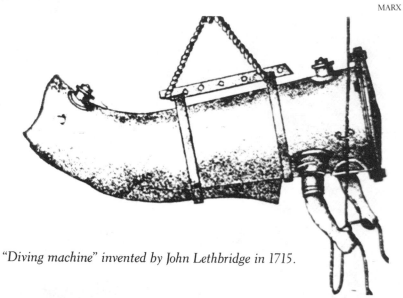

"Diving machine" invented by John Lethbridge in 1715.

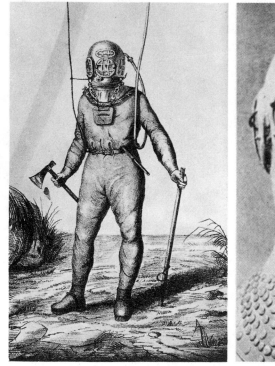

LEFT: *French diver, circa 1815.*
RIGHT: *Helmet of Augustus Siebe's open diving suit, 1819.*

That same year, a Scottish grocer from Edinburgh, Charles Spaulding, invented a way whereby divers could raise or lower a bell at will, independent of any help from the surface. This eliminated the danger of the bell being snagged on underwater obstructions and enabled the divers to maneuver their bell for close-up work on shipwrecks. Spaulding demonstrated the practicality of his bell by salvaging treasure from various shipwrecks in Scotland and England. But in 1783 while working on a shipwreck in Dublin Bay, he and his son lost their lives. Due to strong underwater currents, it was impossible for the casks of fresh air to be lowered to them. Instead of coming up for air when their supply became foul, Spaulding and his son pressed their luck and, as a result, were suffocated by the accumulation of carbon dioxide in their bell. This tragedy led to a change in the way divers in bells were provided with fresh air.

There is also a ghoulish story about an incident that took place in 1790. A team of divers using a Halley diving bell were sent to Senaglio Point in the Bosporus to salvage a shipwreck. Minutes after starting their descent, they signalled frantically to be brought back to the surface. When they were pulled up, the horror-stricken divers refused to go down again. At the bottom they had come on an amazing spectacle: hundreds of life-size dolls in the shape of bowling pins, with skulls for heads. Most of them had been toppled over, but there were rows of them jammed upright among the rocks or stuck in the mud and slime, slightly swaying to and fro in unison whenever the current moved them and, of course, grinning "with a lipless grin," as though beckoning the divers to approach. The divers had stumbled on the spot where generations of concubines from the seraglio had been ritually murdered by being sewn up alive in sacks weighted with stones, with only their heads protruding, and then thrown into the sea at night from boats. The women had either fallen victim to court intrigue or somehow offended the Grand Signor (some of them had as many as 2,000 concubines at one time). Sultan Ibrahim I, who reigned from 1640 to 1648, was said to have drowned his entire harem of 1,200 concubines. The eerie Loreleis that had frightened the divers were their skeletal remains.

The first important recovery of sunken treasure in the New World by anyone other than the Spanish wasn't made for nearly a century and a half after Phips made his. In December 1830 the English frigate, *Thetis,* sailing from Rio de Janeiro to England with $810,-000 in gold and silver bars aboard, was driven by a storm against the rocks off Cape Frio, Brazil and sank in 70 feet of water, drowning half the crew. When the survivors reached Rio and reported the disaster, it was assumed that the treasure was lost forever, for the huge waves and a current of six knots at the site of the wreck made salvage seem impossible.

But Thomas Dickinson, the captain of a British sloop in Rio at the time, decided to go after the treasure despite the risk. He had to contend with more than just turbulent waters, though. There was no such thing as a diving bell in Rio; worse still, there were no divers. Dickinson had rigged a crude diving bell of two iron water-storage tanks by converting a fire-fighting pump into a compressor to provide the bell with air. Arriving at the site, he saw at a glance that, due to the huge seas and strong currents, anchoring within a mile of the wreck was out of the question. So he set up camp onshore and began building a boom on a cliff nearby. Using masts and spars from the *Thetis,* Dickinson constructed a boom 158 feet long and weighing about 40 tons, from which to raise and lower his bell. This project took four months to complete.

Fifty seamen from Dickinson's sloop received on-the-spot training in the art. Most of them had never dived. Fortunately, the water was clear, and during the first month of salvage work, they gleaned $120,000 in treasure. Then a fierce gale sent the boom crashing into the sea. It took Dickinson and his men months to build another one. Battered by 20-foot waves, several bells went the way of the first or were smashed against the rocky cliffs. Dickinson, who was about as ready to yield to defeat as William Phips had been, had his carpenters build others from anything that came to hand. Sometimes they even used wooden barrels. After 18 months of labor and at a cost of three lives, the salvors returned to London, expecting to receive large shares of the $750,000 they had recovered. But they were disappointed. Litigation over the division of the treasure dragged on for years in the Admiralty courts, and finally the salvors were given a one-eighth share to split among themselves, apportioned according to the amount of work each had done.

By the end of the eighteenth century most large European ports had diving bells. They were used for a variety of purposes: constructing piers, bridges, breakwaters, and lighthouses; repairing ships; salvaging shipwrecks; locating good fishing areas and sunken treasure. In some places, when there was nothing else for them to do, divers would take the bells and conduct sightseeing tours. One of the tourists was the Archduke Maximilian who, during a state visit to England in 1818, descended into Plymouth Harbor and picked up a stone on the bottom as a souvenir.

In 1788, in order to lay the foundations of the now-famous Eddystone Lighthouse, situated on a rock off Plymouth which was notorious for claiming many ships over the centuries, John Smeaton, the engineer who had invented the first workable air pump, invented a new type of diving bell called a "caisson." Made of cast iron, it was mammoth compared to the usual diving bell. From six to twelve workmen would descend in it. When they reached the bottom, air would be pumped into the caisson, forcing out the water. The workmen could then go about the job of laying the foundation for the lighthouse. Smeaton and others kept improving on the caissons, increasing their size and the number of people they could accommodate. In 1805 a caisson was lowered to an old shipwreck in Plymouth Harbor where divers worked 10-hour shifts around the clock bringing up 75,000 pounds sterling in treasure and other cargo. The caisson continued to change and by 1840 had lost its familiar inverted-kettle shape. Now it was a massive, elongated, cube of cast iron in which men could spend *days* at a time working. It was used almost universally for building docks, port installations, and deepening harbors. At this time, little or nothing was known about the physiology of diving, however, and many men were killed. From the beginning of long-term caisson diving, the workers of the caissons tended to suffer from vague sensations of discomfort and malaise such as itching skin or pain in their joints. Some were permanently paralyzed, and not a few died painful deaths. In 1850, for instance, eight workmen who were salvaging a ship were killed in the caisson. It became their coffin. Not until 1870 did scientists learn the cause of such deaths—the bends—and caissons became relatively safe to work in. Pneumatic caissons, in which the air was maintained at atmospheric pressure were introduced and soon universally accepted.

4

DIVING MACHINES

The year that ushered in the sixteenth century also began the age of invention in diving. Its herald was Leonardo da Vinci, the Italian genius who was as famous for his inventions as for his paintings. Asked by the Venetians for a breathing device to aid the divers who were taking part in the war against Turkey, Leonardo designed a snorkel breathing tube more advanced than anything seen up to that time. The device consisted of a tube attached to a leather helmet. The entire apparatus fit over the diver's head and had glass windows for seeing underwater. Leonardo also designed swim fins for the hands and feet, which enabled the divers to swim both faster and farther. This gear was the forerunner of the basic skin-diving equipment used by millions of divers today.

But there was a problem. The Venetian Senate rejected the snorkel on the grounds that the Turks would be able to see the breathing tube sticking above water, and the element of surprise would thus be lost. The Senate asked Leonardo to design a breathing tube that would allow divers to approach the enemy in total concealment. So Leonardo designed the first self-contained underwater breathing apparatus, or SCUBA. Although similar to his first design, it differed in two important respects: instead of protruding above water, the tube was attached to a bag made of animal skin and containing air; and to allow the diver to walk on the seafloor, Leonardo discarded the fins in favor of a complete diving suit. Encased in leather from head to toe and with a bag of air on his chest, the diver would descend carrying a heavy weight which he would drop when he wanted to come back up. Leonardo claimed that a diver could remain submerged for four hours.

The design never got past the drawing board. While Leonardo was certain that it would work, he never tested his invention and refused to show it to the Venetian Senate. In his memoirs, he explained his change of heart:

> How and why I do not describe my method of remaining under water, or how long I can stay without eating: and I do not publish or divulge these by reason of the evil nature of man, who would use these as a means of murder at the bottom of the sea, by breaking the bottoms of ships and sinking them altogether with the men in them. And although I will impart other of my diving inventions, in those there is no danger, because the mouth of the breathing-tube is visible above the water supported by bags and corks.

Leonardo's humanitarian scruples were unnecessary, for we know today that his invention could never have worked. Far from staying down the four hours Leonardo claimed, a diver wouldn't have lasted more than a few minutes, and that only in shallow water. If he had descended to even a moderate depth, the pressure of the water would have compressed the bag, thus reducing the supply of air within. Also, by breathing the same air over and over again, the diver risked contaminating his oxygen supply with carbon dioxide, a process that can cause unconsciousness and death. Even if a diver had descended with bags containing enough air to allow him to stay submerged for an appreciable time, and if he were capable of detecting the presence of carbon dioxide (something that isn't easy, as a diver will tell you), he still had another problem: the weights necessary to compensate for the buoyancy created by the air bags would have kept him from moving at all.

A few years after Leonardo's impractical SCUBA (which remained a secret until after his death), an anonymous inventor unveiled his own breathing device. In 1511 three illustrations of divers engaged in military operations appeared in a revised edition of *De Re Militari*, a military work written by the Roman, Flavius Vegetius Renatus in A.D. 375, which had been reprinted many times since and in many languages. The first illustration did not arouse much interest. It showed a naked diver about a foot below the surface of the water, carrying a mace and shield but not using a breathing device. But the second illustration, depicting a diver wearing a SCUBA similar to Leonardo's design, attracted a lot of attention. Whether the inventor arrived at his idea independently or whether Leonardo had shown his design to a friend who betrayed the secret, isn't known. If the anonymous inventor pirated the design, he inherited its disadvantages as well, even adding a few of his own, perhaps for the sake of originality. His helmet had no windows and was therefore useless for anything but underwater blindman's buff. In addition, the air bag was smaller than the one in Leonardo's design; it would have given the diver even less breathing time. Tests probably revealed the impracticality of the design for no more was heard about it or anything like it for a long time.

It was the third illustration that stirred the greatest interest, though today it's difficult to understand why. The device shown was just as unworkable as the second. The diver in the illustration—as ready for blindman's buff as the poor fellow in the second one, but less able to avoid the tag because of his cumbersome diving suit— had to breathe through a tube four feet long, whose open end was kept afloat by two small air bags on the surface. But this couldn't be. The pressure of the water, which increases with every inch of depth, would have made it impossible for the diver to draw air through a tube more than two feet long. In the same way that a snorkel more than two or three feet long is useless, so is a tube of greater length, unless there is a way to pump air through it from the other end.

Amazingly, the busy inventors of the sixteenth and seventeenth centuries don't seem to have realized this important fact. A few years after the periodic publication of *De Re Militari*, another military treatise appeared, which showed a diver breathing through an impossibly long snorkel. This "victim," however, had an advantage over his predecessor: there were windows in his helmet. So he could see out—even if he couldn't breathe. Similar designs followed quickly, all showing tubes or snorkels of fantastic length. Most of them were accompanied by claims alluding to the great depths that could be attained. To the modern diver, it seems incredible that inventors continued to insist on the validity of such claims when any sort of test would have revealed how farfetched they were.

Throughout the sixteenth and seventeenth centuries, devices featuring snorkels and breathing tubes remained in the realm of fantasy. But industrious inventors did achieve some practical success with another device—the diving bell. Like the device described by Aristotle, this one relied on lowering an open, inverted kettle or barrel and holding it level as it descended, to keep the air from escaping. Designed to enclose the diver's head or even his entire body, it often worked.

In 1609 the "Lorini Trumpet" was invented, but there's considerable doubt that it was ever used. If it was used, it would have been a bust, possibly claiming the life of any diver foolish enough to go down in one. The Lorini Trumpet consisted of a reinforced leather tube 30 feet long, which was large enough to permit a man to pass through it. At the bottom of the tube was a suit containing portholes and sleeves into which the diver slipped. Theoretically, the diver breathed air from the surface which he could see by lifting his head. With no way to replenish the air in the tube (which would quickly become foul), the diver would have soon died from carbon dioxide poisoning, not to mention what the water pressure would have done to the leather tube.

In 1628 Diego Ufano, a French military engineer, published a book similar to *De Re Militari*. Ufano described in great detail a diving apparatus which, he claimed, the ancients used and which was currently being used by the French navy.

> In order to facilitate the work of the diver below the surface the Ancients invented a hood most suited to the purpose. It was made of good greased cowhide and so carefully stitched that the water could not enter through any seams. It was so made that it covered the diver from head to middle, and from the shoulders sleeves came down to the elbows, being drawn together and tightened at the extremities in order to prevent the entrance of water. Before the eyes lenses made of transparent horn were carefully let in so that the diver should have uninterrupted vision under the water. On top of the head there was a pipe made of the same leather and stitched with the same care, and this pipe was long enough to extend to the surface when the diver was on the ocean bed. It was held up on the surface by pigs or ox bladders. In this way the man below the surface of the water could be provided with air and enabled to breathe. Down below the diver had weights attached to his feet, in order that the water should not lift him to the surface, and a cord in one hand, the other end of which was attached to some spot on the boat above, so that he might give signals and subsequently be guided back to the boat after having completed his tasks below.

Like many before and after him, Ufano failed to realize the physical principles of water pressure. Consequently, they continued to design equipment that couldn't possibly function. Some of this equipment undoubtedly took the lives of those testing it. A French Jesuit named Père Fournier, who was also a naval chaplain, is regarded as the founder of hydrography. In 1643 he stated that the navies of the Mediterranean were employing large numbers of divers to destroy enemy ships and recover lost cargoes. "These men can remain as long as a quarter of an hour, and even half an hour, under water, just holding their own breath." His statement about the effects of undersea water pressure is even more startling: "Now I say that the reason why a man at the bottom of the sea does not feel the weight of the water which is pressing upon him is because the body which is completely under water is heavier, harder and more solid than the water itself."

It was nearly 40 years before scientists began to suspect the true nature of water pressure. In 1681 Jean de Hautefeuille, the inventor of the spiral spring for clocks, wrote a treatise entitled "The Art of Breathing Under Water" in which he attempted to demonstrate why man cannot breathe air at normal atmospheric pressure when underwater at any depth:

> The first reason is taken from M. Pascal's teaching of the equilibrium of liquids in which he shows in what fashion water acts on submerged bodies pressing in on them from all sides, and in which he demonstrates that a compressible body introduced into water must be compressed inwards towards the center, something he proves by various examples. No great degree of intelligence is required to grasp that a bag or tube attached to the mouth of a man under water will act in the same fashion. A man's lungs are a sort of bellows and in order to inflate them the man must raise the column of water which is pressing down on him from above. However, this column, being very heavy and the strength of his muscles being very inadequate, he finds he cannot raise the column of water and therefore he cannot breathe.

Despite this and other treatises on the subject, inventors kept designing impractical breathing equipment for use underwater. Many even had the audacity to claim that their apparatus did, indeed, work. One such pathological liar was the Frenchman, Benôit de Maillet, who published a book on the subject in 1748. Maillet claims to have employed many divers to help him study the bottom topography of the Levant and Barbary Coast. With the equipment allegedly available, Maillet's divers couldn't possibly have accomplished what he claimed they did.

These divers were equipped with closefitting caps of oilcloth provided with masks and ending in cotton cloth which could be fastened round the neck, that no water could penetrate. Attached to the top of each cap was a long breathing tube of leather which enabled the diver to go down very deep and stay under water for several hours at a time. Each diver carried a compass in one hand and a small pointed stick in the other and at the end of this stick floated a pennon. When such a stick was planted into the sea-bed the diver could readily see the direction and force of the current. These divers were also able to walk around on the bottom of the sea, provided the mud was not too deep.

Maillet's success and his reports of oceanographic research would be more believable if he had used the type of viewing tube (the forerunner of the glassbottom buckets, which is in use today) described in an issue of the *Gentleman's Magazine* in 1782: "A method of discovering anything at the bottom of the sea, which may be useful in viewing the situation of the *Royal George* [a shipwreck]. Take a large tube, like that of a two-foot reflecting telescope (it must be water and air proof) and at the extreme end of the tube fix a clear piece of common window glass, then immerse this tube into the water and apply the eye to the top of the tube and all things underwater will be seen."

While diving bells had their uses, they were also limited. It is noteworthy that there is no mention in the historical records of their use by pearl divers in the New World. If the West Indian and Negro divers could remain submerged for 15 minutes at a time, as Oviedo and others asserted, it's equally obvious that they didn't need bells, particularly. However, there are other reasons for the failure of diving bells to gain acceptance in the pearl industry. An examination of these reasons will help us better understand the limitations of the diving bell. First, it was cumbersome. It took considerable manpower, as well as boats that were much heavier than the dugout canoes the pearl divers ordinarily used. Second, because pearls were often found at depths of about a hundred feet, bells were useless until the advent of Papin's and Halley's bells. But these innovations only made the bells more cumbersome than ever. Third, it was far too costly to provide enough bells for all the divers engaged in the work. Fourth, the need to return to the bells for air restricted the divers' range.

Clearly, the versatility of the bells was limited. Add to this the important consideration that the bells couldn't be used in turbulent water or they would overturn and the air would escape, and it's not surprising that diving bells became unpopular before long. By 1850 they were virtually obsolete. That year marked the beginning of the age of helmet diving, in which air was pumped down to a diver below, who could move about freely. As for reaching depths that no other equipment could reach, the inventors had succeeded in improving the diving bell right out of existence. They had turned it into a new device—the diving chamber—sometimes called the "closed diving bell" or the "diving sphere." The diving chamber was a closed container which could be lowered and raised with cables or weights. In a diving chamber strong enough to withstand undersea pressures, divers were able to reach depths never dreamed of by the early inventors.

The earliest known design for a diving chamber appeared in 1578 in a book called *New and Possible Inventions,* written by an Englishman named William Bourne. Bourne's plan called for a wooden chamber, supposedly watertight, made from the hull of a small ship. Weights were to be used for descent, jettisoning them for surfacing. The time it could remain underwater depended on the number of passengers it carried and how long the air remained pure. The diving chamber was never built. Although its principle was followed in building the first submarine some 40 years later, there was little interest in diving chambers for the next century and a half.

This lack of interest ended in 1715 when another Englishman, John Lethbridge, received a patent for his "diving machine." The name isn't very descriptive, but what else could he have called his invention? Looking like a cross between the old diving chamber and an armored diving suit, it consisted of an irregular metal cylinder six feet long with diameters of two and a half feet at the head and 18 inches at the foot. Armholes in the sides enabled the diver to work with his hands, and a glass window gave him visibility. While the diving machine had no air supply other than the air trapped inside before the chamber was closed, it did have two valves on the top, which could be opened so that air could be pumped in with a bellows whenever the diver had to surface. The device was raised and lowered with cables, but Lethbridge, a man of considerable ingenuity, provided the diver with the means of discarding the extra ballast weights on the outside of the chamber in an emergency and allowing him to ascend unaided. Lethbridge hoped to reach great depths. When he tested his machine, though, he discovered that below 50 feet, the greater water pressure caused leaks around the armholes, window, and entrance hatch. Because of this, Lethbridge's diving machine failed to extend the depths to which divers could descend or the amount of time they could remain below, and the public lost interest in it.

In spite of this, Lethbridge put his machine to good use in the waters of the British Isles and elsewhere in the Atlantic, managing to salvage several valuable cargoes from shipwrecks. In 1724 the Dutch East India Company ship, *Slotter Hooge*, was en route to Java from Holland when it sank during a gale off Porto Santo, on Madiera Island. Of 254 aboard, only 33 survived. In the flooded hold 60 feet below were three tons of silver ingots and three massive chests of coins. Lethbridge, who was considered a technical genius and one of the best salvage divers of his day, was hired by the company to salvage the wreck. His contract specified that he would be paid 10 pounds sterling a month, plus his expenses, with bonuses to be left to the "generosity of the Company directors."

In his first attempt, Lethbridge recovered 349 silver ingots, many coins, and two cannon. Then bad weather forced him to stop. The following summer, he recovered over half of the total listed on the manifest. But after the second year, the amount recovered diminished every year until 1734 when he "came back with less than one could have hoped for, but still with reasonable success." The rest of the treasure was buried in the sand where it would remain for more than two centuries until discovered by a twentieth-century treasure hunter.

About 1750 a diving chamber similar to William Bourne's was built by a cousin of Lethbridge's, a carpenter named Nathaniel Symons. Shaped like an egg and constructed of wood, Symons's chamber used weights to descend and ascend. It wasn't connected to the surface. Although he tested it many times in the River Dart, Symons finally admitted failing to reach any significant depths. Below 50 feet, water pressure would force leaks in the seams.

In 1772 yet another Englishman, John Day, built a diving chamber which also wasn't linked to the surface. Day said that on his first test, he reached a depth of 30 feet and remained submerged for four hours. Historians doubt this, since the air in the chamber, which wasn't large, would have become foul long before then. But at the time, Day's story was believed, and he found a backer in London who advanced him enough money to build what Day called "a proper underwater boat." Decking the hull of a 50-ton sloop to make it watertight, Day announced that he would descend to a depth of 300 feet and remain there for 24 hours without communicating with anyone on the surface. Fearing that this was a bit foolhardy, his backer persuaded Day to take the chamber down only to 130 feet and ascend after 12 hours.

Several thousand spectators were at Plymouth Harbor, June 29, 1774 to witness the descent. The chamber descended into the harbor on schedule. The 12 hours passed, and then 24 hours. It wasn't until this point in the demonstration that there was alarm. The crowd merely assumed that Day had reverted to his original plan. Grappling hooks were used in an attempt to pull the chamber to the surface, but no trace of it or its inventor was ever found. As might be expected, there was much speculation about what had happened to Day and the chamber—everything from Day having suffered frostbite to the diving chamber being devoured by a sea monster. What probably happened was that he died from inhaling foul air or the water pressure opened the seams and he drowned.

In 1774 an article appeared in the *Gazette de France* about a curious diving event:

> On March 19th the King of Sweden, accompanied by Princess Sophie-Albertine and a numerous suite, went to Lake Wartan to examine a new machine invented by a Mr. Hammer for the use of divers. The king accorded him the privilege of recovering nautical effects from the bottom of the sea. Mr. Jonas Dalhberg, Inspector of Divers, descended to the bottom of the lake where he walked around over a space of 120 feet in area in a way that permitted the King and the princess to see him walking. He then sang several verses of song which his majesty heard very clearly by means of a tube seventy feet long. After that he sang another song for the princess, who heard it by means of the same device.

Unfortunately, this is all we know about Mr. Hammer's "Machine."

Then in 1798 a Mr. Kleingert of Breslau, Germany designed and built a new machine which he described as the "ultimate diving apparatus." Consisting of a cylinder of strong tin plate, which enclosed a diver's head and body, it was constructed in two separate pieces to facilitate putting it on. In addition, the diver wore a jacket with short sleeves and a pair of drawers of strong leather, all watertight, "and joined by brass hoops round the metal on the outside, so that he was relieved from the pressure on all parts, except the legs and arms. With this apparatus, on the 24th of June, 1798, a man named Joachim, under his direction and before many spectators, dived and sawed through the trunk of a large tree at the bottom of the River Oder."

The next attempt took place in 1831 when a Spaniard named Cervo built a small wooden diving chamber. Cervo claimed that he could reach a depth of 600 feet with it. On the first trial, in which he was to dive to 200 feet, Cervo failed to come up. This time, we know why: pieces of the chamber floated to the surface, indicating that it had been crushed by the pressure. Cervo's was the last recorded attempt to reach great depths in wooden diving chambers. After this, inventors began to look for materials more durable than wood.

Two Americans, Richards and Wolcott, designed a metal diving chamber in 1849. Globular in shape, it was meant to be lowered and raised by chains connected to the surface. But there was no provision for an air supply. Still, they might have succeeded had it not been for the lack of money, which prevented its construction and testing. News of Richards' and Wolcott's idea of using metal spread quickly, and inventors all over the world turned to building metal diving chambers.

The first to come up with a workable one was a French engineer named Ernest Bazin who built his chamber in 1865 for use in the search for a treasure sunk in Vigo Bay, Spain. Made of high-grade steel and connected to the surface by cable, it had two viewing ports and an external electric lamp for underwater illumination. Like the Richards-Wolcott chamber, Bazin's didn't have a supply of fresh air other than what was trapped before submersion, but this did not keep it from reaching 245 feet, almost three times the existing record, where it remained for an hour and a half. The problem of supplying the diver with fresh air was solved by a Venetian named Toselli who in 1875 improved on Bazin's design by providing the diving chamber with a large cylinder of compressed air which was capable of sustaining the diver for up to 50 hours underwater.

The way was now open. Diving chambers continued to be improved until they were capable of reaching depths which such pioneers as de Lorena, Papin, and Halley could only have dreamed of. Inventors went on building chambers of thicker and stronger metals, making them safer and more comfortable for divers, and adding countless improvements such as cameras and radios. In 1934 two Americans, Dr. William Beebe and Otis Barton, descended in a diving chamber called a bathysphere. This diving chamber functioned according to the same principle as did Bazin's chamber; it had simply evolved further. This bathysphere descended to a depth of 3,028 feet, a record that was broken over and over again in the years to come.

There were limits even to what a bathysphere could do, though. No winch in the world was strong enough to lower and raise a diving chamber that heavy to depths significantly lower. Clearly, something else was needed, something far beyond a mere improved design. The man who came up with it was a Swiss physicist, Auguste Piccard whose invention, the bathy*scaphe,* could lower and raise itself by means of weights—the principle of Bourne's diving chamber. In 1953 the bathyscaphe reached a depth of 10,395 feet, and in recent years, others have been built which have dived even farther, culminating, in 1960, in the dive of the *Trieste* to 35,800 feet, the deepest part of the ocean known. One of the two men in this record-breaking dive was Jacques Piccard, the son of Auguste Piccard.

It must not be assumed that dives such as these are made merely to break records. In such an age of scientific curiosity as ours, man's desire to know all he can about the world he lives in is the primary motivation for attempting to reach the seafloor. While bathyscaphes have helped widen our knowledge of a strange environment, they have other uses. For instance, in 1963 the *Trieste* located and photographed the nuclear submarine *Thresher* which had been lost on a test dive in 8,400 feet of water off Massachusetts.

5

DIVING WITH A HELMET

For centuries the possibility of walking freely on the seafloor clad in a diving suit and breathing air from the surface through tubes seemed as remote as walking on the moon. Then in 1715 an English inventor named Becker demonstrated his new invention in London. The device consisted of a full, leather diving suit and large, spherical metal helmet with a window. Three tubes led from the helmet to the surface, one for exhaled air and the other two for fresh air pumped down by several large bellows. When it was demonstrated, the diver stayed down for an hour. Unfortunately, the depth he reached wasn't recorded. That Becker's invention was put to practical use, we know from the eye-witness account of a traveler visiting England in 1754. This observer saw divers trying to salvage a recently lost warship. They were supplied with air by means of a pump rather than by the primitive and unreliable bellows.

During the eighteenth century the system of pumping air from the surface down to divers seems to have been used only in England, for there's no record of its being used elsewhere. While the British were in the forefront of undersea technology, inventors on the Continent were still turning out drawings and designs which showed divers trying to suck air down through long tubes, as the illustration in *De Re Militari* had shown two centuries earlier. Because British pumps could provide sufficient air to divers working only 15 to 20 feet below the surface, perhaps it isn't surprising that the rest of the world took no notice. Not until the invention of the air compressor, which permitted air to be forced down to divers at greater depths, did the kind of diving known as helmet diving become popular.

In 1771 a Parisian named Freminet introduced his "hydrostatergatic machine." The diving rig consisted of a helmet made of brass, with three small portholes, one of which was near the mouth. This hole was for a copper tube which connected to "a kind of oval object, rather like a small iron rugby football nine inches long, which the diver casually tucked under one arm." Another copper tube connected the other end of the air cylinder to the back of the helmet. This device, the forerunner of today's SCUBA, contained air at atmospheric pressure, which meant that the diver was limited to only a few minutes on the bottom. Breathing air over and over again for a longer period would have resulted in asphyxiation and death. The helmet was connected to a leather suit covering the diver from neck to feet, which was reinforced with an articulated metal framework.

After moving the air cylinder to the diver's back, Freminet claimed that in 1776 he descended 50 feet to the bottom of the Seine, where he stayed for 30 minutes. On another dive in the harbor at Brest, he reportedly spent an hour on the bottom, nailing lead sheathing to a ship's bottom and raising an anchor. There must be a special deity who protects such inventors, for a 10-minute dive in Freminet's rig would probably kill you today!

Several documents that have survived in the French National Archives tell of successful diving experiments conducted in Toulon Roads, but they don't describe the diving apparatus made by a naval official named Sardou. They do state that Sardou's "hydraulic machine" enabled him to make three dives, for a combined time of 22 minutes, and that with the exception of his legs, which weren't covered, he stayed dry. "To the amazement of all present he returned from each submersion with stones and other debris he found on the harbor bottom."

Also in 1776 a short article appeared in a London newspaper, concerning an experimental diving suit built by an unidentified Englishman. "At Liverpool a person attempted to go down in a diving apparatus to the wreck of the *Pelican*. He descended about four fathoms and a half; but owing to one of the tubes breaking, and wont of proper persons to work the air pumps, was obliged to be taken up immediately to prevent suffocation."

And in 1819 another Englishman—Augustus Siebe, sometimes called the father of helmet diving—invented the diving suit and helmet that has evolved into the standard helmet diving rig in use today. Siebe's invention consisted of a brass helmet into which air was pumped from the surface by hand-operated air compressors and a suit of leather, with a canvas overlay to keep the suit from being cut underwater. Both the excess air pumped down to the diver and his exhaled breath were forced out through vents at the waist. But the suit, which was known as an "open" diving suit because of the vents, had a major disadvantage: the diver had to remain upright at all times, or the air pumped into his helmet would rush out through the waist vents.

Still another Englishman, this one named Caston, almost suffocated on dry land while wearing a helmet diving rig. In 1836 Caston, who owned a junk shop, somehow came into possession of the equipment.

> Mr. Caston determined to try it in the first instance on terra firma and for this purpose drew the helmet over his head, and then adjusted that part which fitted the lower extremities. He however omitted the most essential part of the apparatus - namely, the valve which admitted the air into that portion which fitted over his head and face. This neglect nearly cost him his life; for when one of his servants entered the warehouse, Mr. Caston was discovered rolling on the floor, enveloped in the diving apparatus, apparently in great agony and the servant saved his master's life by extricating him from the horrendous apparatus.

Two English brothers, John and Charles Deane, who were in the marine salvage business, obtained patents in 1823 on the basic design for a "smoke apparatus" that would permit firemen to move about in burning buildings. By 1828 this had evolved into "Deane's Patent Diving Dress," consisting of a heavy suit for protection against the cold, a helmet with viewing ports, and hose connections for bringing in air from the surface. The helmet wasn't fastened to the suit. It simply rested on the diver's shoulders, held in place by gravity and straps attached to a waist belt. Exhausted or surplus air passed out from under the helmet, and this posed the same problem that Siebe had encountered: if the diver tripped and fell, the helmet would quickly fill with water. In 1836 the Deane brothers issued the first diver's manual and made a small fortune from their salvage business.

In 1834 an American, L. Norcross, improved on Siebe's "open" diving suit by placing an exhaust vent for excess stale air on top of the helmet and thus eliminating the waist vents. Divers could now bend over and even lie down underwater without cutting off their air supply.

Probably not wanting to be outdone by an American, Augustus Siebe modified his open diving suit in 1837, making it watertight and incorporating Norcross's exhaust vent. But Siebe went further. His exhaust vent was designed so that it could be controlled by the diver who expelled air only as necessary. There were two advantages to providing him with the means of controlling his air supply. Air pressure could be built up within the suit when the diver wanted to descend; this way, he could resist the increasing water pressure which would otherwise press the suit tightly enough against his body to stop his circulation and cause temporary paralysis. Also, in the event that the diver couldn't communicate with the surface and be pulled up, he could inflate his suit with air until it filled up like a balloon and surface under his own power. This safety factor has saved the lives of many divers over the years.

The first opportunity for Siebe to prove the worth of his *closed* diving suit came soon after he had tested it and put it on the market. Fifty years earlier, the 108-gun Royal George, the largest and most important warship in the British navy, had sprung a leak and sunk at its anchorage at Spithead (off the southern coast of England), with the loss of a thousand lives. The wreck was a serious menace to the port, at that time, the most important in Great Britain, and for years, attempt after attempt was made to raise it. Although the water was only 65 feet deep, every attempt failed. Winches weren't strong enough to lift so large a ship. The Admiralty then tried to blow it up, but the delayed action fuses on the explosive charges failed to go off. In 1834 the Admiralty asked Charles Deane to survey the wreck and assess the possibility of raising it. During the next three years Deane made many trips to the site. Wearing the Siebe open diving suit, he recovered many items, including 30 large cannon. Deane eventually came to the conclusion that there was no way to bring the *Royal George* up in one piece, since marine worms had eaten through much of the wood.

Thoroughly frustrated, the Admiralty launched a campaign to get rid of the wreck at any cost. It even started a diving school for the Royal Engineers who, not coincidentally, were experts on explosives.

Divers working on the wreck of the Royal George.

Wreck of the Royal George.

LEFT: *Alexander Lambert in the treasure chamber of the* Alfonso XII.
RIGHT: *Diver digging for gold ingots on the wreck of the* Laurentic.

The man in charge was a colonel named Pasley, an expert diver himself. Colonel Pasley had no difficulty training his men, but that was all that was easy for him. After testing Siebe's open diving suit, he came to the conclusion that the suit was too dangerous and refused to allow its use. He attempted to use a diving bell but found this impossible due to the strong tide. In fact, no equipment then in existence would work.

This was where matters stood when Siebe announced the successful testing of his closed diving suit (soon to be known as the "standard helmet diving suit"). Colonel Pasley, now ready to try anything, found it ideal for the job. Divers placed watertight metal cylinders filled with dynamite at selected points on the wreck. Then, when they were out of the water, the dynamite was fired electrically. Later they returned to the wreck to clear off the debris and prepare for further demolition. The explosions had to be small because the surrounding waters were always crowded with ships. The job took over five years, but at last it was done, thanks to Augustus Siebe and his diving suit. A new technology—underwater salvage and demolition—had been born.

Miraculously, there had been no fatalities among the divers working on the wreck of the *Royal George*. But there were quite a few accidents. The most serious occurred when a diver's air hose broke and the air instantly rushed out of his helmet. He said afterward that he felt as though he were being crushed by the water. Luckily, those on the surface saw what had happened and pulled him up immediately. His face and neck were swollen, and he was bleeding from the ears, eyes, and mouth. That diver spent several weeks in the hospital. He never dived again.

Nevertheless, the diver was fortunate. He had survived the first recorded case of the "squeeze," a disaster feared more than any other because it's nearly always fatal. The squeeze occurs whenever there is a sudden drop in the air pressure inside a suit and helmet. If it happens in deep water, the enormous pressure pulls the diver's skin off his bones and jams it into his helmet. (Once the flesh of a Japanese diver was not only forced into his helmet but through his air hose as well.) Today the danger has been reduced considerably by installing a safety valve to prevent the diver's air from escaping.

In 1838 the British Society of Arts presented a silver metal to a Mr. Thornwaite who had invented a special diving belt that enabled divers to surface without assistance from the surface. Several divers had lost their lives using Auguste Siebe's helmet rig, which had been invented the year before, and another safety device was badly needed. Thornwaite's belt was made of India rubber cloth to which was attached a small air reservoir with compressed air equivalent to 30 or 40 atmospheres. When the diver wanted to ascend, he merely opened a valve which inflated the belt and sent the diver to the surface. But tests showed that the ascents were too rapid, resulting in serious accidents and sometimes death. Divers smashed against diving boats and succumbed to the "bends" and other "diving ill-nesses," about which very little was then known.

An inventor named E. Heinke received a similar award in 1851 for his improvements on the diving helmet. "Among the most prom-inent of the improvements is that of a double valve fixed in front of the gorget, which enabled the diver to descend and rise at pleas-ure, with the whole of his gear, which weighs upwards of 200 pounds; in fact, it places the whole apparatus completely under his control, and protects him in case of anything happening to the air-hose, as by its means a sufficient quantity of air to support respira-tion for ten minutes can be contained in the helmet and dress, thus giving time to ascend even from a very great depth."

The European manufacturers of helmet diving gear competed vigorously. During 1855 two interesting trials were held in England and France. Siebe had again modified his helmet, including im-provements made by Heinke, and the Admiralty decided that it was far superior to the others. But tests in the Seine at Paris resulted in the French and some other governments present adopting Heinke's equipment. Three British and two French "diving-dress" and hel-mets were tested. "In order to test the alertness of the divers, twelve small rings were thrown into the river; of these Heinke's divers picked up ten, and the other two were not found by any of the divers." Heinke received another award, this one from the French government. Siebe, however, soon regained his prestige by having one of his divers perform a feat thought impossible in the dark, cold, muddy waters of the Thames, near London. A woman thief who was being hotly pursued by the police threw some valuable watches off Blackfriars Bridge. It was no easy job to find the watches in the muddy bottom, since the gold in them would cause them to sink into the mud. Despite this, Siebe's diver found all 10 watches, to great acclaim by the press.

In 1859 a man making his first dive wrote an interesting account of his experience. He used Seibe's diving rig to descend in the Thames.

This vessel, firmly moored, was, as it were, the base of their operations. There it was that the divers took up their quarters before going down into the water, or when they came up again from the bottom of the sea. There also, on a kind of platform, stood the air-pumps and the men appointed to work them. These air-pumps appear on the outside to be nothing but mere boxes, in shape like a large travelling-trunk, but containing on the inside steel cylinders, and a complete system of ingenious mechanism. When it is wished to set this mechanism in motion, they fix outside the box a fly-wheel and two crank-handles, which are turned round by two workmen by manual power. Instantaneously a jet of air escapes with great force through a valve opened in the lower part of the pump-case.

Just at the time when I got out of my boat the divers were on board, and were taking their hour of rest. As I was bent on obtaining every information both as to the nature and use of their equipment, I asked if they would allow me to put on the diving-dress. This they consented to. A description of this apparel, every article of which forms part of a system, can hardly be considered extraneous in an account of modern submarine invention. The professional divers are clothed entirely in woolen material, and when they have to descend into very deep water, they also protect some parts of their body with a basket-work enclosure covered with rough flannel. All these under-clothing arrangements are nothing more than precautionary sanitary measures. They soon brought me the actual diving-dress: it is a large grey garment, made of india-rubber, all in one piece, and of course waterproof. One has to get into it from above, as if into a sack, and it is finished off by a pair of trousers with feet, as well as two sleeves. The upper part of this garment was then closely fastened round my neck by a handkerchief, and round my wrists were placed india-rubber rings, so as to fasten down on the flesh the ends of the sleeves, already tight enough, as I thought. I had not much difficulty in understanding that this latter precaution was intended to prevent any water getting in. I was then shod with a heavy pair of shoes with leaden soles, each weighing ten pounds. [After they] put a woolen cap on my head, my shoulders were loaded with the metallic helmet collar - a kind of pelerine of tin, polished like steel, with a copper edging. By means of holes bored in this edging, and screws which fitted into them with wonderful accuracy, the lower part of the collar is hermetically fixed on a leathern pad which runs round the top of the dress, encircling the chest and back.

I was thus encased, when a friendly hand placed on my head a helmet of spherical shape, with a large glass eye on each side, and one opening only, opposite the mouth. It was through this I breathed, and certainly not over and above well. My position reminded me somewhat of the story of the Iron Mask, especially when they riveted the helmet firmly on to the metallic collar of which I have just spoken. To add to my accoutrements, a belt was placed round my loins, to which was hung a dagger-knife, the formidable blade which was enclosed in a copper sheath, the only means for defending it from contact with the water. The dagger-knife was intended to sever the gordian knots which I might meet with on my subaqueous journey, and no doubt also to give me means of defence against any marine monsters I might fall in with. All I now wanted was an axe in my hand, and then I should have resembled a real diver - as much, at least, as a caricature resembles a portrait. But no, I was in error. The divers called my attention to the fact that at present I was too light, and that as I was I should not be able to descend to the bottom of the sea. They therefore hung on, in front of my chest and behind my back, two pieces of lead weighing forty pounds each. Now, at all events, my toilet was complete. When I looked at the shadow which I threw on the deck of the vessel, I could hardly help laughing. What fabulous animal could it be into which I was metamorphosed?

But it was not my shadow only which astonished me. Whenever I tried to speak, my voice sounded hollow and dull in the cavity of the helmet. I asked the divers whether, being already so far advanced on the road, I might not attempt to descend to the bottom of the Channel in good earnest. They stared at me with wondering eyes, in which I fancied I read a feeling of doubt and anxiety. "At all events," said one of them, "he can do as he likes."

This same diver undertook to give me the requisite instructions. He showed me the way to get rid of, by one movement of the hand, the two pieces of lead which were fastened on my breast and back; and assured me that, when this was done, I should immediately rise up to the surface. "I should not give this advice," he added, "to a diver by trade; for we consider it rather a disgrace to leave our lead weights at the bottom of the sea, having other means of calling for help in case of danger or accident. The diver is in communication with the surface," he continued, "both by the air-pipe and by a rope which we call the life-line. Both of these appliances speak a language of their own. But of all these signals there is only one which concerns you: if you wish to come up, you must give four pulls to the air-pipe, and that means haul up; it will not be long before your wish will be understood and complied with. The whole time that a diver remains under water, there are on board two reliable men, whom we call attendants, who watch over him just as a careful nurse does over a child walking with leading-strings. These three men may, in fact, be said to form but one, and the whole science of diving is based on this system of perfect concurrence. The two attendants are forbidden by our rules to talk to each other, or even to any other person, during the whole time that they are performing their duties. Talking might divert their attention, and they need to use the utmost vigilance to catch the sense of the slightest signal. They are, in fact, answerable for the life of the man at the bottom of the sea. Now then," he added, "you must consider if your heart misgives you, or if you think you can enjoy a short time all alone along with the waves. Oh! I had almost forgotten an important piece of advice. Sometimes it happens that the diver loses his way in the sea, and finds himself unable to regain the bottom of the ladder by which he went down. To guide us in a perplexity like this, we employ a cord rolled round the wrist, which unrolls as we walk along. If, from any special cause, this clue happens to fail us, the diver may insure his safety by making the signal of distress - Haul up! - and he is immediately taken out of his difficulty by being drawn up to the surface.

I assured the diver that all this latter advice might be very good, but that, as far as I myself was concerned, it was perfectly useless, for I had not the slightest desire to venture very far in a region which was so perfectly unknown to me. "Well, I half suspected as much," replied he, smiling.

The helmet which covered my face and head was provided on the back part with two hollow metallic studs: one of these was protected against the intrusion of the water by a strong valve, and was intended to give vent to the air vitiated by breathing; the other, called the pipe-holder, was to be fixed to the air-tube. I had noticed, in fact, on the deck of the vessel, a long india-rubber pipe, coiled round and round like a slender serpent. One of the sailors took hold of the head, as it were, of this elongated reptile, and screwed it into the air-pump, whilst he inserted the other end - the tail, so to speak - in the pipe-holder or metallic stud on my helmet. I could then well understand how the whole theory of this art is based, as might be expected, on the physical constitution of man. The diving-apparatus only doubles and lengthens his respiratory organs; the air-pump is for him nothing but his external lungs, and the air-tube is only a floating wind-pipe. It was not long, however, before they closed up the only orifice by which I had any communication with the outer world, by screwing on, in front of my mouth, a third pane of glass, which was oval, and protected, like the two others, by a copper wire-work. We must not lose sight of the fact that there is nothing but a thickness of glass between the diver and the ocean - between life and death. If any external obstacle - some projecting point of iron - should chance to break this frail barrier, in a moment he would have to deal with all the waters of the deep.

Almost before they had finished fixing on this glass in front of my helmet the pumps began to work, and to supply me with air; but for this I should have been stifled, for the only part of my person which was now in contact with the atmosphere were my hands, and I could not effect much in the way of breathing through them. This function was now entirely dependent on the air-tube; but oh, if this tube happened to break! It was explained to me that even under these unlikely circumstances, a valve would close spontaneously to prevent the rushing in of the water, and that enough air would be left inside the diving-dress to enable me to live some instants, just time enough, in fact, for me to be rescued. This was at least some consolation. I could now neither speak nor hear, but I could still see very well - for had I not my three glass eyes? I was directed by signs to make my way to a ladder which went down into the sea from the side of the vessel, but the difficulty was for me to move. I seemed as if I was glued down to the deck by my leaden-shoe-soles; my back and breast were loaded with weights, and besides, I felt as stiff and uncomfortable in my india-rubber garment as if I had been sewn up in the skin of some marine monster. However, I did my best, and at last reached the first round of the rope-ladder, which was stretched pretty tight by a considerable weight at the lower end, and passing round the side of the ship above the water, then disappeared.

The good-hearted sailors, however, did their best to help me, and guided all my movements; they taught me how to pass the air-tube under my left arm, and the signal-line, bound round my body, passed upwards over my right shoulder. The upper ends of the tube and line were held by two men, who were, in divers' phrase, my attendants. I do not reckon a third man who accompanied me to show me the way. The ladder seemed a very long one to me, although there were not more than eight or ten feet between the edge of the vessel and the level of the sea; but the terrible moment is that when one touches the surface of the waves: although the sea was that day as calm as a pond, I felt myself beat about and buoyed up by the natural movement of the waves rolling one over the other, in spite of the leaden weights attached to me. But it was much worse when my head went under the surface, and I saw the water dancing about round my helmet. Had I too great a supply of air in the apparatus, or had I too little? Really it would be difficult for me to say; the fact is, that I felt almost suffocated. At the same time, it seemed as if a tempest was roaring in my ears, and as if my temples were screwed up tight in a vice. In good truth I had the strongest desire to go up again immediately, but shame was more powerful than fear, so I slowly descended - too slowly for my liking - for this ladder down into the deep appeared as if it would never end, and yet the water at this spot was not more than thirty or thirty-two feet deep. I could scarcely summon up presence of mind enough to observe the gradual deterioration of the light round me; it was a pale, doubtful twilight, which to me very much resembled the London atmosphere in a November fog. I fancied I saw living forms floating about here and there, without at all being able to say what they were. At last, after a few minutes, which seemed to me a whole century of trial and torment, I found my feet were resting on a surface which was something like solid. My reason for thus modifying my expression is that the bottom of the sea, when you are on it, does not appear to be a very satisfactory base: every moment one is buoyed up and half carried off the legs by the moving masses of water, and I was compelled to hold tight on to the ladder with my hands to prevent being tumbled over.

There was, however, one essential instrument which I was in need of: the divers, to enable them to walk firmly in the ocean, are in the habit of using a crowbar, on which they lean as on a walking-stick, but I was quite encumbered enough already without this iron bar, which would have been no use to me. For I had not the slightest intention of walking about; I was much too dismayed at the impressive silence and gloomy solitude of the water, in which I seemed to myself lost. Here, however, the light was much brighter than it appeared to be when I was halfway down, and the pains in my head left me as if by magic. Wishing to carry back with me a tangible proof of where I had been, and a souvenir of my excursion, I stooped down and picked up a pebble from the bottom of the sea. I was going to put it in the pocket of my dress, when I found out that I had no pocket, and that I must stick it under my girdle.

This being done, I gave the signal to hoist me up to the surface.

By 1860 the helmet diving rig had replaced the diving bell and even many of the caissons used in submarine architectural work such as laying the foundations of piers, seawalls, and breakwaters. Companies sprang up in every major port, where helmet divers were employed to clean and repair ship bottoms. About this time there was a resurgence of interest in salvaging old shipwrecks. Many treasures once thought lost forever were recovered. In just one year of salvaging, a diver received over £20,000 in bonuses and retired. Another diver told of a melancholy scene when he was sent down to explore the wreck of the *Dalhousie,* a ship that had sunk off the coast of Scotland. When he entered the main cabin, he "found a mother on her knees in an attitude of prayer, and clasping her two little ones in her arms, whilst the other dead bodies were clinging on by their finger nails to the beams of the ceiling." Another helmet diver exploring a wreck off Ireland found a beautiful young woman lying peaceably in one of the berths, her long, disheveled blond hair floated about like seaweed with the movement of the water. "I took good care," he said, "not to disturb her in her sleep: where could she have found a more peaceful tomb?"

In 1860 the *Malabar* sank in the English Channel where for many months she lay untouched in a hundred feet of water. Such a depth was thought impossible to tackle then. After Lloyd's of London wrote off the ship, a team of divers from Plymouth, using Heinke's diving rig, took on what the newspapers described as a "suicidal project, full of sheer folly and extreme recklessness." But within three days the divers salvaged the entire treasure worth £280,000—at a price: two of them died from the bends.

Another major treasure was recovered in 1885 from the Spanish ship, *Alfonso XII,* which had sunk off Point Grando on Grand Canary on her way to Cuba. *Alfonso XII,* which was carrying a half-million dollars in gold coin, lay under 162 feet of water and, like the *Thetis,* was believed lost for good. To add to the difficulty, the treasure was stored in a strong room on a lower deck, and there were three iron decks between it and the top deck. The insurance company was so sure that salvaging her was hopeless that they paid the owners the $500,000. But they hadn't reckoned with Alexander Lambert who had already distinguished himself by twice closing the iron door of the flooded Severn Tunnel against formidable odds. Because no other diver would descend almost twice as far as it was considered safe for helmet divers to go, Lambert went down alone. In just two days, by blasting his way through the three iron decks and breaking into the strong room, he recovered the entire treasure. Lambert, however, paid a penalty, that was becoming only too common. He got a severe case of the bends and was forced to retire from diving permanently.

The greatest treasure recovery in history—and probably the most difficult ever undertaken—was one sponsored by the British government. Early in 1917 the converted White Star liner, *Laurentic*, now an armed cruiser, sailed from Liverpool for Nova Scotia, carrying in its second-class baggage room gold bars valued at $25 million. While still within sight of the Irish coast, she hit a German mine, exploded, and sank in 132 feet of water, with a loss of 300 lives. Salvage would be formidable, although treasure had been recovered from greater depths and diving techniques had been improved considerably since Lambert's achievement in 1885. There was also the danger that a salvage vessel would be sunk by a German U-boat or a floating mine. Nevertheless, the attempt had to be made. The loss of so much gold would deal a serious blow to a British economy already weakened by the war. The sinking of the *Laurentic* had to be kept secret from the Germans, who almost surely would have tried to block recovery efforts. For reasons of security, the British public wasn't informed either.

Captain G.C.C. Damant, a naval salvage expert, was put in charge. A few weeks after the disaster, his salvage vessel, disguised as a fishing trawler, appeared on the scene, carrying some of Britain's best helmet divers. Upon inspection, the divers found that, because of the 60-degree angle at which the wreck lay, walking on or inside it would be impossible. They would have to crawl, pulling themselves along by grasping whatever projections they could find. The route to the second-class baggage room, where the gold bullion lay, was blocked by a watertight door halfway down the ship's side, which had to be blasted open. After this was done and the debris had been cleared away, the divers discovered that the passage beyond was filled with floating wreckage which also had to be cleared away. This job alone took hundreds of diving hours. The next obstacle was another door, which was also blasted open. At last they reached the door to the baggage room, and a diver opened it with a hammer and chisel. He carried out one of the boxes of gold, which weighed 140 pounds and was worth $40,000. The weight wasn't as great as it would have been on land, but owing to the difficulty of crawling along the slanted passage, he could retrieve only one box that first day. The next day they took out three more.

In good weather the operation might have proceeded well, but the salvage site was battered by a succession of winter gales. Due to rough seas, the salvage vessel was hard pressed to maintain its position over the wreck, and the divers' hoses were constantly in danger of being snapped. Then, shortly after the fourth box of gold had been raised, a gale of such force sprang up that Captain Damant was forced to order the trawler to seek shelter in the nearest port.

They returned to the site as soon as the storm had moderated. But the first divers to go down brought back discouraging news: the *Laurentic* had collapsed like an accordion, and settled even deeper into the sediment on the seafloor. The entry port on the side, which had been at a depth of 62 feet before, was now 103 feet below the surface. Worst of all, the passage beyond the door, originally nine feet high, had been compressed to a mere 18 inches, too narrow for the divers to get through. Difficult enough before, the job had taken on monumental proportions. Even if they could get through the constricted passage, because of the greater depth at which they would have to work, each diver would now be able to spend only 30 minutes below and 30 more decompressing. The work was fatiguing; no diver could make more than one descent a day, and only two could work at the same time. Otherwise, air lines might become entangled. Not surprisingly, the salvaging progressed at a snail's pace. Explosives were used to blast a path through the wreckage. Finally, after the dangerous job of clearing the debris had been done, the divers once again reached the baggage room. It was a shambles. The floor was full of holes, and there was no gold to be seen. The boxes containing it had disintegrated along with everything else. So the gold bars must have slid through the holes, deeper into the wreck.

Captain Damant needed a new plan, for the wreck was crumbling with every passing hour. For instance, the decks above the baggage room were being supported only by water. Damant decided to cut straight down through the decks to the spot where he judged that the gold had fallen. As usual, "cutting" meant more demolition. After some two months of blasting, gold bars were sighted in the part of the wreck that had sunk into the seafloor. Divers then groped for the bars. This slow, grinding work went on day after day until September when bad weather again forced Captain Damant to suspend the operation for the year. They had recovered $4 million in gold so far.

Because Captain Damant and his team were needed for another highly secret salvage job, there was no work at the *Laurentic* site in 1918. When the salvors returned to the *Laurentic* in the spring of 1919, they found only a pile of wreckage which bore little resemblance to a ship, not even what had once been a ship. Again the divers cleared a path to the spot where they had found gold before. This time they found a total of $2,350,000. But then came a period in which they could find no more gold. Damant concluded that the remainder must have fallen into a depression deep within the tons of wreckage. Until this debris was cleared away, the divers could not continue salvaging. Throughout the summers of 1920 and 1921 the divers cleared away debris, using large centrifugal pumps to remove the tons of mud by pumping it to the surface. Only 50 bars of gold turned up during these two years, but Captain Damant and his men were determined to continue looking.

Then in 1922, their luck changed. The first diver down that spring sighted several gold bars sticking out of the mud on the seafloor. For once the winter storms had worked in their favor, washing away a lot of the silt that had been covering the bottom. Between April and October, when work had to be stopped for that year, they found about $7,500,000. The most that was found in a single trip was $750,000. The salvors did even better in 1923, recovering nearly $10 million. There were now only 154 bars, valued at $1,200,000, remaining to be recovered, and during 1924, 129 of them were recovered. By this time, it had become unprofitable to continue the effort, and the operation was abandoned. It had been a highly successful venture for the British government, which got it all, since the salvors were members of the British navy and received no share of what was recovered. Ninety-nine percent of the bullion aboard the *Laurentic* had been recovered, at a cost of only 2 percent of the amount recovered.

Because of the large number of vessels sunk during the American Civil War, the Merritt-Chapman & Scott Company of New York was founded. The company soon became the largest, most prestigious salvage firm in the world. Over the years, divers were involved in salvaging, repairing, and refloating hundreds of ships. One of the company's major salvage operations took place in 1918 when the 13,000-ton liner, *St. Paul*, sank in New York Harbor in 54 feet of water. The *St. Paul* had been converted into an armed cruiser during World War I, which meant that the first thing the divers had to do was remove her munitions. More time was spent removing the other moveable objects. To lighten the ship as much as possible, the superstructure was blasted away. Cofferdams were built along the deck, and all of the openings were sealed off before air was pumped into the wreck. In addition, the divers had the dangerous task of digging tunnels beneath the ship, in order to sling cables and put lifting pontoons in place. Six months later, the 22 divers had done their work well. As the wreck would have been a hazard to navigation in New York Harbor if left there, the *St. Paul* was refloated and towed away to be cut up for scrap.

During the Second World War Merritt-Chapman & Scott had over two dozen salvage vessels operating around the world. Serving as a civilian arm of the U.S. Navy Salvage Service, the firm received 498 major assignments. All told, their helmet divers reclaimed ships and cargoes valued at approximately $675 million, which is merely a statistic that doesn't begin to tell the whole story. The *Official History of the Navy Salvage Service* said it better: "No mere figures in dollars and cents can adequately measure the value of the recovered hulls and cargoes to the successful prosecution of the war."

Around the turn of the century, large oyster beds were discovered off the coast of western Florida. An enterprising American went to the Aegean Islands where he talked some 200 Greek helmet divers and their families into emigrating to Tarpon Springs, Florida. Since then, the port has become the world center for sponges. It is still the Greeks—handing down their skills from father to son—who control this lucrative business. Many of them still wear helmets and diving dress manufactured in the nineteenth century, to the great delight of the thousands of bug-eyed tourists who come to Tarpon Springs each year.

MAN GOES DEEPER

As the use of the standard helmet diving suit became more wide-spread and improved air compressors allowed divers to reach greater depths, it was discovered that a diver who used the suit risked perils other than the "squeeze." One of the most serious of these was and is the "bends," so named because of the agonizing contortions the diver who has them goes through. When he reaches extreme depths, the air pressure within his helmet and suit must be equal to the water pressure outside or he will be crushed. For example, at a depth of 200 feet, the diver must receive air pressure seven times that of the atmosphere on the surface. Since air is 21 percent oxygen and 79 percent nitrogen, there is an excessive amount of both gases in his air supply; how excessive depends on how long he remains below. Oxygen—which is used up by the body tissues, exhaled in the form of carbon dioxide, and released through the exhaust vent—presents no problem. But nitrogen does. It remains inert until the diver ascends to a lesser depth, at which time it comes out of solution and is expelled by the lungs. The problem arises when he ascends too quickly. Then the excess nitrogen forms bubbles in the bloodstream and blocks the blood's circulation, resulting in either temporary, but usually permanent, paralysis or death. Many divers have died from the bends.

The actual cause of the bends, or "caisson disease," was first clinically described in 1878 by Paul Bert, French physiologist. In his study of the effect of pressure on the human body, Bert learned that breathing air under pressure forces quantities of nitrogen into solution in the blood and tissues of the body, resulting in the bends. He recommended that caisson workers decompress gradually and that divers return to the surface slowly. Bert's research led to an immediate improvement in the working conditions of caisson workers when they found out that their pain could be relieved merely by returning to the pressure of the caisson as soon as the symptom appeared. In a few years, there were specially designed "recompression chambers" at job sites, to provide a better environment for controlling the bends. Pressure inside the caissons could be increased or decreased as necessary for a particular worker. One of the earliest uses of the recompression chamber was in the construction of the Holland Tunnel under the Hudson River in 1893, connecting New Jersey with New York City. Use of the chamber dramatically reduced the number of serious accidents in such construction work, as well as reducing significantly the number of deaths from the bends.

Bert recommended that divers ascend gradually and steadily, but some divers didn't take this seriously and some continued to suffer from the bends. It was widely believed at the time that divers had reached the practical limits of the art, that 120 feet was about as deep as anyone would ever be able to work effectively. This was due partly to the frequent cases of divers getting the bends from deep dives and partly to the marked inefficiency of the diver working below 120 feet. Divers often lost consciousness below this depth.

From 1905 to 1907, J. S. Haldane, a British physiologist, tested many divers and found that there was a fairly simple reason for the bends: divers weren't ventilating their helmets properly. As a result, carbon dioxide was building up to dangerously high levels. The problem was solved by establishing a standard flow rate of 1.5 cubic feet of air per minute and providing pumps with a capacity to maintain this flow. Now the helmet was adequately and continually ventilated, which made the diver's work safer. An immediate result of Haldane's research was the extension of the effective operating depth for helmet divers to about 200 feet. This limit wasn't a physiologically imposed one but rather one imposed by the ability of existing hand pumps to provide air. Helmet divers still had to contend with the bends.

Eventually, as the problem of the bends came to be more highly recognized, research on the condition was begun, both in England and the United States. Investigators learned that if a diver ascends slowly, stopping at successive levels to allow the excess nitrogen to dissipate, he can avoid the bends. This process is known as *decompression*. Diving tables, indicating how long a diver must remain at various depths, were compiled and are still used today. While decompression will prevent the bends, the process is a long, tedious one. A diver who descends to 300 feet and remains for an hour must, in his ascent, spend 4 minutes at 120 feet, 10 minutes every 10 feet until he reaches 70 feet, 14 minutes at 60 feet, 28 minutes at 50 feet, 32 at 40 feet, 50 at 30 feet, 90 at 20 feet, and finally 187 minutes at 10 feet before surfacing—a total of more than seven and a half hours of decompression after an hour's work underwater. The deeper the diver goes and the longer he stays down, the longer he must spend decompressing. For this reason, decompression is far from the ideal solution to the bends, and scientists are still looking for a better one.

Ever since the standard helmet diving suit became a reality, making it possible for divers to move about freely underwater, divers so equipped have been assigned most of the undersea tasks. The British navy, consistently the leader in the field, has used helmet-suit divers to inspect and repair the bottoms of ships. The United States, however, didn't begin to use them until the late 1870s. A few years later, in 1882, the first helmet diving school was begun at Newport, Rhode Island. For many years thereafter, however, there was little official interest in this new branch of naval operations. Divers were given a mere two weeks of training, hampered by regulations which limited diving depths to 60 feet, and usually assigned to the lowly job of recovering spent torpedoes. Meanwhile, British divers were working in depths of up to 130 feet. Thanks to their pioneering efforts, much has been learned about the perils that await divers in deep waters.

In 1898, when the *U.S.S. Maine* was sunk in Havana Harbor, American divers got their first chance to show what they could do. They recovered the battleship's cipher code and keys to the munitions magazine, thus preventing them from falling into enemy hands. Such diving skill should have opened the navy's eyes to the tremendous potential of diving, but after routine congratulations were offered, the navy forgot about the divers and they slipped into obscurity. Then in 1912, Chief Gunner George Stillson, a seaman in the U.S. Navy, wrote a report for the War Department in which he was highly critical of the navy for allowing its diving program to lag so far behind that of other countries. Unaccountably, where performance had failed to stir the navy brass, a report sent through channels succeeded, and in 1913 Stillson was ordered to set up an experimental diving school at the Brooklyn Navy Yard, which would replace the original one in Newport.

With only a handful of divers, Stillson began experimenting in reaching greater depths, using the newly tabulated diving tables. In 1914 he himself set a depth record for the open sea by descending to 274 feet. The achievement, which received worldwide attention, resulted in Congress authorizing the money for more research in diving. The Experimental Diving School was moved back to Newport, and the next year, divers trained by Stillson reached the submarine *U.S.S. F-4* which had sunk near Honolulu, at a depth of 304 feet, breaking their teacher's depth record in the process.

The staff and students of the school were needed for salvage work in Europe during World War I. When the war ended, however, Congress refused to authorize funds for reopening the Experimental Diving School. That this was shortsighted economy was proved when, in 1925, the submarine *U.S.S. S-51* sank off Block Island in 132 feet of water. Although Stillson's divers had reached 300 feet 10 years before, only about a dozen divers were now qualified to dive beyond 90 feet. Consequently, only three of the *S-51*'s 37-man crew were saved, and it took more than a month to raise and salvage the submarine. Many people became aware of the need to have qualified divers on standby for emergencies. But Congress still refused to authorize the money. Then in 1927, there was another disaster. The submarine *U.S.S. S-4* sank in 102 feet of water with a loss of 40 crewmen. The salvage operation took three months.

Spurred on by the press, people were indignant over the navy's failure to save the men trapped in the *S-51* and *S-4*. The furor forced Congress to pass legislation funding another Experimental Diving School. The school was located in Washington, D.C., where it is today. Some of its courses were designed to develop better diving techniques and reduce the risk of nitrogen narcosis, using the newly discovered helium-oxygen mixtures, while other, more specialized courses were for perfecting methods for rescuing men trapped in sunken submarines and providing the means for them to escape on their own if there was no immediate help.

The invention of a submarine escape chamber, or "McCann

chamber," so named after its designer, made the rescue of men trapped in submarines less hit or miss. The chamber, which resembled an ordinary diving chamber, was lowered directly over the escape hatch of a sunken submarine. Then great pressure was built up inside it to keep water from rushing in when the door was opened. Before the hatch of the submarine was opened, some of the men to be rescued stepped into the room just beyond the hatch, and the air pressure was increased there. Then both doors were opened, the men entered the McCann chamber, both doors were shut, and the rescued men were taken to safety. Meanwhile, the air pressure in the room beyond the escape hatch was reduced and a new group of men stepped in to await rescue. When the McCann chamber returned from the surface, this procedure was repeated, until all the men trapped in the submarine had been rescued. Tests showed that the McCann chamber would be invaluable in rescue operations, and the divers of the Experimental Diving School became proficient at using it.

In 1939 they got the opportunity to demonstrate their proficiency. On May 23 the submarine *U.S.S. Squalus* sank off the Isle of Shoals in the north Atlantic. The next morning, the diving vessel *Falcon* located the *Squalus* in 243 feet of water, and a helmet diver was sent down to attach a steel cable to the conning tower so there would be no chance of losing it and to rap on the steel hull to let the men inside know that help was on the way. During the next 12 hours the McCann chamber made four round trips, taking all 33 survivors from the forward section of the submarine to safety. The chamber was then moved to the aft escape hatch, but this section of the sub had been flooded and there were no signs of life. An effort was made to raise the submarine, which involved many divers and hours of labor. Thanks to the helium-oxygen breathing mixture and careful adherence to the diving tables, there were no fatalities among the rescuers. The rescue operation was a complete success. The *Squalus,* raised and later recommissioned as the *Sailfish,* saw action in World War II.

This incident convinced the navy that divers were here to stay. Because there were too few of them, and with war looming, the navy pressed Congress to allocate more money for training divers on a large scale. Congress appropriated the money. About this time, the former French liner *Normandie,* which had been commissioned for service in the U.S. Navy, caught fire and capsized at her mooring in New York Harbor. A new diving school was set up right there on Pier 88 to give the student divers on-the-spot training in salvage techniques. The school remained there until the *Normandie* was raised in 1946 when it was moved to Bayonne, New Jersey. Then in 1957 it was incorporated with the Experimental Diving School in Washington. Divers had become a respected part of the U.S. Navy.

Although the United States finally committed itself to a thorough diver-training program, until recently, British helmet divers were far ahead in salvage techniques. Moreover, from 1900 to 1940 there were 10 trained helmet divers in Great Britain for every diver in the U.S. The reason isn't hard to find. Out of necessity, Great Britain, an island nation, has not allowed herself to fall behind in any aspect of maritime technology. This meant that funds for diver training and experimentation were always forthcoming from Parliament.

Shortly after World War I it became apparent that the money hadn't been wasted. On June 21, 1919 the defeated Germans scuttled their entire fleet inside Scapa Flow (which had been the main British naval base during the war), to block the harbor and prevent their ships from falling into British hands. One of the ships scuttled was the 25,000-ton battleship, *Kaiser,* the largest ship in the German navy, but there were others almost as large. With the base virtually inaccessible to their ships, the British found themselves in a difficult position. They rushed every available helmet diver in the British Isles, whether Royal Navy or civilian, to Scapa Flow, and the greatest salvage effort up to that time began. Hundreds of divers were employed in the project. Without them, the port might have been denied to England for decades to come. As it was, the last bit of debris wasn't removed until just before World War II.

About the time that research on the bends was beginning, it was discovered that the heavy concentration of nitrogen in helmet rigs could cause trouble even before divers ascended. The problem came to light when it was noticed that divers were behaving strangely in deep water. Some chased fish, while others turned somersaults or did imitations of Nijinsky. At first, employers thought maybe they should have had their divers tested for sanity, but eventually, when reports of these antics had become more persistent and widespread, the cause was identified as nitrogen narcosis, or "raptures of the deep." Below 200 feet, and sometimes before that, the accumulation of nitrogen in the air supply of divers was having a narcotic effect on them, so much so that they would lose all sense of danger. Death often resulted when divers, unaware of what they were doing, cut their air lines or removed vital parts of their equipment.

Even before the submarine disasters in 1925 and 1927, there had been concern that navy divers would be called on to perform rescue and salvage work in very deep water, much deeper than a helmet diver breathing compressed air could work at, either intelligently or safely. In 1919 Professor Elihu Thompson, an electronics engineer and inventor, suggested using helium in a diver's breathing mixture, instead of nitrogen. He went to the Bureau of Mines which, at that time, was trying to find a use for helium, now available in sufficient quantities. After surmounting numerous obstacles, such as lack of money and interest in his idea, in late 1924 Thompson finally convinced the U.S. Navy to underwrite his research. Early experiments indicated that the helium-oxygen mixture had several advantages over the mixture normally used in deep dives. Besides eliminating the narcosis induced by too much nitrogen, the helium-oxygen mixture also seemed to cut down on decompression time, thus making it possible for divers to work longer on the bottom. By early 1927 Thompson's work had progressed so well that the navy moved his laboratory from Pittsburgh to their Experimental Diving School in Washington, and in 1937 a navy diver, using the mixture, reached a simulated depth of 500 feet in a special diving tank at the school. This achievement proved beyond any doubt that depth was no longer an obstacle to submarine rescue and deep-water salvage work.

For foreign divers, however, they might as well never have heard of the helium mixture. The United States possessed the world's only supply of helium, a supply so limited that very little could be exported. Scientists began a search for other combinations. During World War II a Swedish engineer named Anne Zetterstrom substituted hydrogen for nitrogen in the breathing mixture (a substitute that did not free divers from the need to decompress any more than the helium-oxygen mixture had). Zetterstrom went back to his experiments—dangerous experiments, for hydrogen and oxygen are highly explosive when combined. After several years of work, mainly to find the proper ratio of gases, he made his first big test. Wearing the standard helmet diving suit, Zetterstrom descended to 363 feet, and on August 7, 1945, he reached a record-breaking 528 feet, 88 feet deeper than American divers, using the helium-oxygen mixture, had gone. Unfortunately, the dive cost Zetterstrom his life. An assistant on the surface accidentally shut off the oxygen in his air supply, and he died from breathing pure hydrogen. Although the hydrogen-oxygen mixture worked, it was considered too dangerous for general use. But what effectively relegated the hydrogen-oxygen mixture to the category of a scientific curiosity was the discovery of helium in the United States, where it occurs as a minor constituent of natural gas. Restrictions on its export were lifted, and the mixture has become standard.

The British were also experimenting with various gas mixtures. After numerous experiments in tank-simulated dives, they were ready for the real test: a deep dive in the sea under the same conditions divers face when called on to perform at great depths. Lieutenant-Commander J. N. Bathurst, who was in charge of the project, chose the submarine salvage vessel *Reclaim* as the surface support vessel and the calm waters off the port of Trabert, Scotland, as the dive site. Using a helmet diving rig that had been used by hundreds of other test divers, William Bollard began his well-planned dive at 9:21 on the morning of August 14, 1948. A minute and a half later he announced over his telephone that he had reached 180 feet and was switching over from regular compressed air to the special helium-oxygen mixture. For the next 45 minutes, Bollard "denitrogenized" himself at this depth until the medical experts on the surface were satisfied that it was safe to continue deeper. At 360 feet they had him stop again. To test his alertness and physical condition, they had him perform a number of mechanical chores and compute some mathematical equations. Then the dive was resumed. When Bollard reached the record depth of 540 feet, he again performed some physical tasks, simulating a diver taking part in a submarine rescue and helping with a salvage project.

"What are your impressions?" he was asked. "I could go deeper, but my fingers feel dead." For five minutes he hung at this depth before asking to be brought up.

Bollard now faced the most dangerous part of the dive—the ascent and decompression. At 260 feet he was halted for seven minutes and then again at 210 feet for a longer period. At 180 feet he entered a large diving chamber where assistants removed his helmet and checked his physical condition. Then the chamber was slowly raised, with stops every 10 feet. Each time Bollard's condition was checked for signs of the bends or any other problems. At 60 feet his breathing mixture was switched to pure oxygen. He said he was fine but very weak. It had been three hours and ten minutes since Bollard had first entered the water. At 30 feet an assistant in the chamber suddenly yelled: "Quick, quick, haul up! He's having an epileptic fit, he's writhing with pain!"

When the chamber had been taken aboard, the doctors discovered that the compressed oxygen was affecting Bollard's nervous system. They quickly put him in a decompression chamber where he was given the helium-oxygen mixture again under pressure corresponding to a depth of 80 feet. For three hours and twenty minutes the pressure was gradually reduced until it reached one atmosphere. Bollard was alive but completely exhausted, and it took him several weeks to recover. The dive had been successful, but it was obvious to scientists and deep-diving experts that much more research was needed before deep diving would be safe.

By 1958, when a Swiss mathematics teacher named Hannes Keller began work on an alternative gas mixture, researchers had come to the conclusion that a dive to the depth Bollard had reached would require a decompression period of several *days*—which was obviously impractical. Keller believed he could solve the problem by combining several different gases. The implications were enormous. By freeing divers from the long periods of decompression, such a mixture, if found, would enable them to reach greater depths and spend more time there. Keller, also an amateur SCUBA diver, sought the help of Albert Buhlman, who was a physiologist and cardiac specialist at the University of Zurich. Together they combined nine gases. Using a computer, Keller calculated the complicated tables giving the proportions of the gases at each stage of a diver's descent and return to the surface.

In 1959 Keller made his first trial dive. Standing on a platform in a lake, surrounded by cylinders filled with his mixture and breathing through a regulator like that on an aqualung, he descended to 400 feet—platform, cylinders, and all—and surfaced without pausing for decompression. There were no ill effects. Elated by his success, Keller went to France where he tried to get financial backing from

Captain Jacques-Yves Cousteau, a man famous for his underwater experiments. Although skeptical, Cousteau arranged a demonstration at the Undersea Research Group Laboratory in Toulon, France, in an experimental recompression chamber which simulates the atmospheric conditions of an actual dive. The gas mixture was pumped in at pressure corresponding to that of water at 630 feet. When Keller dived and surfaced without any decompression, Cousteau was impressed but not convinced. He though Keller was some sort of freak with an extraordinary tolerance for gases under high pressure, and refused to support him.

Next, Keller went to the United States and talked to the navy. He met with skepticism there, too, receiving no help then or a promise of any in the future. So he went back to Toulon. Using Cousteau's recompression chamber, Keller made a simulated dive to 1,000 feet and surfaced without decompression. He underwent extensive medical tests which indicated that he had suffered no ill effects. This should have convinced everyone, but it didn't. Part of the reason was Keller's secrecy about the composition of his mixture. Which wasn't surprising. The mixture wasn't patented, and Keller didn't want it pirated.

Upon returning home to Switzerland almost penniless he decided the only way to convince the skeptics was to take someone else on a dive with him. If it succeeded, no one could say the reason was his own tolerance for gases under high pressure. Then, perhaps, there would be money for further experimentation. The man he selected for the dive was Kenneth MacLeish, an editor at *Life* magazine. Diving in Lake Maggiore, on the Italian-Swiss border, Keller and MacLeish reached the fantastic depth of 725 feet, returning to the surface without decompressing and in excellent condition. The experiment made headlines, but better yet, the dive silenced the skeptics. Many who had previously turned Keller down now offered him financial assistance, including the U.S. Navy.

In 1962 Keller and Peter Small, a London newspaperman who was also an amateur SCUBA diver, descended to 1,000 feet off Catalina Island, California in a diving chamber of Keller's design, the *Atlantis*. The dive was viewed on closed-circuit television. Unlike other diving chambers, which maintained air at surface pressure, the *Atlantis* contained a mixture of gases at a pressure slightly more than that of the water at the chamber's current depth. This was necessary because it was part of Keller's plan to open a hatch and leave the chamber once he reached the bottom. Unless the pressure inside was at or near the pressure outside, water would rush into the chamber. The descent went like clockwork. Keller left the *Atlantis*, breathing his gas mixture through a hose and an aqualung regulator, and planted Swiss and American flags on the seafloor. Then he returned to the *Atlantis* and bolted the hatch. Unknown to Keller, however, one of his fins had caught in the hatch port. The precious breathing mixture so carefully calculated by the computer was escaping. As the assistants on the surface watched their television screens in horror, Keller and Small lost consciousness. The *Atlantis* was quickly pulled up to 200 feet, where it was met by Richard Anderson, a professional SCUBA diver, who connected an air hose.

Keller regained consciousness but Small didn't, not even when Keller gave him artificial respiration. Small was dead by the time the *Atlantis* reached the surface. Although it was a valuable experiment, the dive cost not only Small's life but the life of Chris Whittaker, another SCUBA diver who had accompanied Anderson on the rescue dive.

7

ARMORED DIVING SUITS

Another way a diver can reach great depths without having to decompress is with the armored diving suit, which differs from the helmet diving suit in that it's constructed of steel to withstand great water pressure. Thus the diver inside breathes air at normal surface pressure. He doesn't have to worry about excess nitrogen, helium, or whatever. In fact, he's subject to the same conditions as someone who descends in a diving chamber or even a bathyscaphe.

The first armored diving suit appeared in 1838, a year after Siebe had introduced his closed diving suit. Equipped with articulated arms and legs, it was the invention of W. H. Taylor, an Englishman. At a time when helmet divers were working at depths of less than 100 feet, Taylor's suit was designed to reach 150 feet. Within a decade others were being made to reach depths of 200 feet, and by 1900 some could even reach 300 feet. As the bends and nitrogen narcosis came to be recognized as a serious threat to helmet divers, interest in the armored diving suit grew.

The armored diving suit never became widely used, due to several drawbacks. The most serious was that their weight and bulk prevented mobility underwater. They must depend on surface attendants to move them around on the seafloor. Another was the expense of manufacturing and maintaining them. Still another was that if they developed leaks, water would rush in, drowning the divers. Armored suits have been used mainly for work at great depths (those in existence today can penetrate as far as 2,000 feet) or at shallower depths when long periods of time are involved.

In 1882 two Frenchmen, the Carmagnale brothers, built and

Salvage of the sunken submarine U.S.S. Squalus.

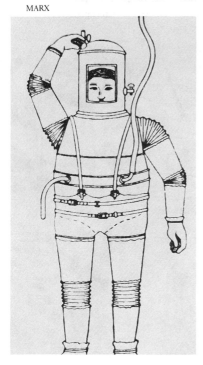

LEFT: *Taylor's armored suit.*
RIGHT: *American armored diving suit, circa 1920. It was used in the recovery of almost two million dollars over a ten-year period.*

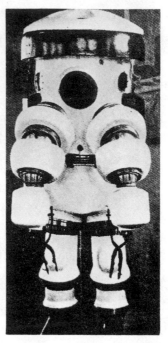

German armored diving suit, circa 1920.

LEFT: *Observation chamber being used on a deep shipwreck.*
RIGHT: *Quaglia's chief diver stands beside the diving observation chamber used in the salvage of the* Egypt.

patented an armored suit which gave the diver more mobility. On earlier models, the accordion-type joints shrank and stiffened under pressure. The Carmagnales' robot-like suit incorporated 22 ball-and-socket joints which were covered with linen for further waterproofing. At the front of the helmet were 20 small viewing ports. The complete rig weighed over 600 pounds. No one knows whether it was ever tested underwater or what depths it could reach.

Soon thereafter, an Englishman named William Carey designed a suit that used roller and ball bearings to facilitate the movements of the joints; but this design wasn't tested for many years. Meanwhile, other inventors worked to perfect more flexible suits. Some were stiffened with spiral wires in an effort to achieve greater mobility than that of the rigid, all-metal suits. But none of these designs worked.

In 1913 the German firm of Neufeldt and Kuhnke designed a more mobile and flexible suit which it put in production in 1920. Combining ball bearings and ball-and-socket joints, the suit required less metal. Also, about this time, an Italian named Galeazzi and an Englishman, J. S. Peress, made a few improvements in the overall design of the armored suit.

Ever since the *Lusitania* was sunk by a German submarine off the coast of Ireland in 1915, the inventive mind has been busy trying to figure a way to get at it. The wreck lies in an upright position at a depth of 42 fathoms, or 252 feet. It is commonly believed that, in addition to several million dollars worth of copper ingots and other valuable cargo, there is another six million dollars in gold, jewelry, and securities in the purser's safe and strongroom. But the depth was beyond the safe working depth for helmet divers at the time, and a fast-running tide made underwater work even more hazardous. Several attempts by British divers to survey the wreckage almost cost them their lives.

Among the many proposals received by the Admiralty, that of an American, Benjamin Franklin Leavitt of Traverse City, Michigan, seemed to have the most merit. In an armored diving suit of his own design, Leavitt had already descended to 361 feet in Lake Michigan, a greater depth than the *Lusitania* now lay at.

The design of Leavitt's suit differed significantly from that of its predecessors. He used manganese bronze, an alloy that won't rust but which has the tensile strength of steel. The complete suit weighed only 350 pounds. Due to the displacement of water, this weight was reduced to 75 pounds underwater. The arms and legs, which could be bent at the elbows and knees, were of flexible tubing. The lining of the suit was sheet-rubber insulation, and the diver could wear any amount of additional clothing. Finally, the helmet had four half-inch-thick shatterproof glass windows and was equipped with a telephone.

But the most striking aspect of this suit was that it was self-contained. The diver did not breathe air pumped down from the surface but rather from a steel tank containing oxygen. The tank was attached to the back of the suit and was connected to the diver's helmet by a flexible copper tube that passed through a regenerative attachment. This attachment, in turn, consisted of a case, or canister, which was filled with caustic soda, for filtering impurities from the oxygen. This meant that the oxygen could be breathed over and over again. A gauge on the tank automatically regulated the flow of oxygen, to compensate for the oxygen consumed by the body. Depending on a diver's individual metabolism, the oxygen supply lasted from three quarters of an hour to a full hour. The diver was lowered and raised by means of a nontwisting steel cable which he controlled by communicating with the surface attendants on the telephone. Although with some effort, he could walk on the bottom, his lateral motion was generally controlled by swinging the derrick or boom from which he was suspended.

Leavitt outfitted the 2,800-ton vessel, *Blakeley*, with heavy lifting equipment, grab buckets, powerful electrical lights, and six of his armor suits. When the permission was not forthcoming from the British to salvage the *Lusitania*, Leavitt went after the *S.S. Pewabic* which had sunk in Lake Huron in 1865 at a depth of 182 feet. Several helmet divers had already died at this wreck. During the summers of 1917 and 1918 Leavitt salvaged 350 tons of copper ore, using his armored suit to direct the movements of the grab buckets.

In early 1923 Leavitt moved to the coast of Chile where the British brig *Cape Horn* had sunk in 1859. The brig, which was carrying a cargo of copper ingots, had sunk in 220 feet of water. Although there had been several attempts to find her, Leavitt, using the technique of dragging a steel cable between two ships, found the *Cape Horn* in the record time of two weeks. Then he settled down to salvaging the wreck. The first few weeks were spent blasting open the wreck. Then tons of wheat, stored in bags, had to be removed. It took a month to reach the copper ingots and salvage them, which netted Leavitt a profit of over $300,000 on the operation.

Leavitt returned to New York, hoping to sell stock in his proposed *Lusitania* salvage project which, he said "would make all prospective backers millionaires." Just before Leavitt reached the $100,000 mark, which was the amount he said he needed, the British Admiralty announced that a hoax had been perpetrated, that the *Lusitania* had not been carrying any gold bullion. In fact, the Admiralty said, the total amount in the purser's safe was probably worth less than £50,000. This claim, coming from such an authoritative source, brought Leavitt's fund-raising to a halt temporarily. But he let it be known that there were still enough copper ingots and other valuable cargo to make the project lucrative. But the British government refused to grant him permission, and the project was dropped. Leavitt continued to demonstrate his armored suit in other profitable salvage operations.

No more was heard of the *Lusitania* until 1935 when divers Gene and John Craig received worldwide attention by announcing plans to descend to the wreck and photograph it. But this project, too, was blocked by the British who simply refused to grant the Craigs permission. The divers had conducted a number of experiments with Dr. Edgar End of Marquette University. Using an armored suit of Max Gene Nohl's design and a mixture of helium and oxygen, they thought they could reach the wreck in five minutes, spend two hours, and make only three decompression stops of 10 minutes each on the way up. With his design, Nohl had set a new record of 420 feet in Lake Michigan. His armored suit was used successfully to salvage ships which had sunk in deep water in the Great Lakes, as well as off the west coast of South America.

Another famous salvage effort involved the Pacific & Orient liner, *Egypt*. On a dark foggy night in 1921 the *Egypt*, en route to Bombay with a cargo of six million dollars in gold, collided with the French cargo steamer *Seine* in the Bay of Biscay. Within minutes, the *Egypt* sank in over 400 feet of water. Because of the depth, salvage experts said it would be impossible to recover the treasure, and Lloyd's of London paid the owners' insurance claim.

But there are always people willing to tackle the impossible, particularly when a lot of money is at stake. The challenger this time was an Italian, Giovanni Quaglia, director of the Sorima Salvage Company of Genoa. After signing a contract with Lloyd's of London and the original owners to divide the profits, Quaglia made his plans. Helmet divers couldn't be used because of the wreck's depth (helium-oxygen mixtures had not yet been developed). First he thought of using an armored diving suit with articulated arms and legs, then decided on a "diving observation chamber." The diver inside could direct the entire salvage operation, and they would use explosive charges and huge electromagnets to remove metal from the wreck. A scissors grab would be used to lift the treasure.

Quaglia tested his equipment on two other wrecks before tackling the *Egypt*. The tests were successful and profitable. In 1928 Quaglia hired the captain of the *Seine* to help him find the wreck. The search went on for three years, with bad weather sometimes holding things up for months at a time. (While waiting for the weather to improve, Quaglia would work on other wrecks in more sheltered areas, thereby increasing the efficiency of his team, while at the same time adding to his profits. One of the wrecks that he raised was the Belgian ship, *Elizabethville*, which carried eight tons of ivory.) Two salvage vessels, the *Artiglio* and the *Rostro*, combed the area where the *Egypt* had sunk, dragging a steel cable over the bottom. Each time the dragline snagged on something, the diving observation chamber was lowered so that divers could see what the obstruction was. Sometimes it was another wreck, but more often it proved to be only a rock on the seabed. Finally, in August 1930, the *Egypt* was located in 426 feet of water.

The following morning, blasting began. They would have to blast their way through three decks that were above the strongroom. The chamber was lowered, followed by an explosive charge on a cable. The diver in the chamber showed the man handling the winch where to place the charge. Then the chamber was raised and the charge was detonated. The chamber was again lowered, this time to direct the electromagnets and scissors grab in removing the debris. This procedure was repeated hundreds of times, with the weather causing frequent interruptions.

When no work could be done on the wreck, Quaglia followed his usual practice of moving the salvage vessels to other wrecks. One of these sideline activities, however, led to disaster. It involved the

Florence H, a United States munitions ship that had sunk in Qui-
beron Bay, France, in 1917. The belief was that the munitions
aboard couldn't explode after so many years in the water—but they
could. The charges planted by Quaglia's team to destroy the wreck
set them off. The *Artiglio* was blown to bits, killing 13 men and
destroying most of the salvage equipment.

Disheartened but not defeated, Quaglia returned to Genoa and
fitted out the *Artiglio II.* He returned to the *Egypt* in 1931, and
blasting was resumed. They reached the strongroom late one evening that autumn. The diver in the observation chamber enthusiastically reported that the floor was "paved with gold bars." But early
the next morning, before any of the gold could be brought up, a
major storm blew up, and it was six months before Quaglia and his
crew could resume their work.

Returning the next spring, they were dismayed to find that the
hole they had blasted through the decks had filled up with debris
again. It was a month's work to clear it. Then more bad weather set
in. Quaglia returned that summer and was finally rewarded for his
efforts. By the end of October, when operations for the year ended,
they had recovered over $2 million in gold bullion, and the next year
most of the remaining $4 million was recovered. Performed in
deeper waters than salvors had ever gone before, the venture was a
success at last.

Within a few years, Quaglia's record was broken. In 1940 the
Australian mail steamer *Niagara,* carrying $12 million in gold, hit a
mine 30 miles from the entrance to Whangarei Harbor, New Zea-
land where the water was 438 feet deep. Thanks to the deep-sea
salvage methods pioneered by Quaglia, the Australian divers who
were conducting the dive didn't need to break new ground. They
also had the advantage of some echo-sounding equipment that had
been developed for use in antisubmarine warfare. But even with the
echo-sounder, the actual search for the *Niagara* was the most difficult part of the job. The wreck lay right in the middle of a heavily
mined area, and on several occasions the salvage vessel was almost
blown up. To further complicate matters, Japanese submarines
lurked nearby.

Since the wreck lay on its side at a 70-degree angle, a diagonal path
to the strongroom would have to be blasted. Using the same type
of equipment that Quaglia's team had used, the Australians com-

pleted the job in nine months. Miraculously, the ship remained intact during the blasting. By comparison, getting out the gold was easy. All but about 6 percent of the gold was recovered. The remainder was thought to have fallen through a hole deeper into the wreck as a result of the blasting. When the salvors heard that the Japanese had got wind of the operation, they decided not to risk going after the rest.

In recent years the search for offshore oil deposits has made it necessary to go deeper and deeper. Diving technology has not been able to keep up, and this has resulted in something of a revival for the armored diving suit. Oceaneering Incorporated, one of the major firms providing diving services to the offshore petroleum industry, has designed and built an armored diving suit named "JIM." With it, a diver can work at depths up to 1,500 feet. JIM proved itself in a series of test dives undertaken in April 1976 off Melville Island in the Canadian Arctic. It was used for a total of 15 hours' work on an oil rig at over 900 feet. Diver Walter Thompson set a world record by spending six hours on the rig in 915 feet of frigid water. This new design is an important factor in offshore drilling technology. The suit will likely exert a strong influence on the engineering decisions of the industry.

There are numerous advantages in using JIM. The diver can descend and ascend at any desired rate. He can reach a thousand feet in only five minutes and return to the surface in the same length of time. He can stay down for long periods, the length of time depending on the stamina of the diver/operator. Little surface support equipment is required. A small half-ton winch and stand are enough to lower and lift the unit. No decompression time is needed, regardless of the operator's bottom time or depth. JIM is highly mobile on a good surface, and the suit weight underwater can be varied for special jobs. In case of trouble, the operator releases weights and JIM floats to the surface. The limitations of the unit aren't much of a hindrance. JIM performs most tasks a bit slower than a diver using a helmet suit or SCUBA equipment. The diver must rely on his vision, which could be difficult in murky water. Finally, the unit has limited vertical movement and cannot function well while dangling in water.

The first commercial application of JIM was in the recovery of some valuable anchor chains lost by tankers moored in a harbor in the Canary Islands. It was shown that the unit is not only valuable as a working tool beyond normal working depths, but is also an excellent search tool. On this particular job, divers using JIM searched 156,000 square yards of the ocean floor before finding the chains in 375 feet of water and attaching lines to them so they could be raised. On another occasion, with winter storms in the North Sea, JIM divers spent a total of 88 hours in 26 dives to a depth of 384 feet, making videotapes of the base of a loading tower for closer inspection on the surface by scientists.

By the time this book goes to the printer, JIM is expected to have worked at its maximum depth of 1,500 feet. Oceaneering Incorporated is planning to build several more JIMs in the near future, and a more sophisticated model, called "SAM," is on the drawing boards. It's expected to have an operational depth of 3,000 feet. At the present time, Oceaneering has a manned, one-atmosphere, self-propelled observation/manipulator bell capable of working in 3,000 feet of water. The firm is also designing an unmanned, deep-water television/manipulator vehicle which will operate by remote control from the surface to reach twice this depth.

SCUBAs

As a result of the great amount of publicity he has received, many people think Captain Jacques-Yves Cousteau is the inventor of the self-contained underwater breathing apparatus, known as "SCUBA." Actually, Cousteau is only one of a long line of men who developed and perfected the SCUBA. In 1869, when Jules Verne's *Twenty Thousand Leagues Under the Sea* was published, many divers were already using SCUBA equipment designed and manufactured four years earlier by two Frenchmen, Benoît Rouquayrol and Auguste Denayrouze.

The desire for total concealment underwater, which had prompted the Venetians to reject Leonardo da Vinci's snorkel in 1500, motivated later inventors to devise versions of a self-contained underwater breathing apparatus that would allow divers to approach enemy shipping without any telltale reeds or tubes. These inventions were no more workable than those of Leonardo or the anonymous inventor who publicized his work in *De Re Militari*. Inventors continued to work at it, however, no doubt because they knew that success would be highly remunerative.

In Spain, for nearly 300 years the most war-like nation in Europe, inventors were especially busy. Spanish archives are full of early SCUBA designs, some with possibilities, some ludicrous. When Dutch ships were driving Spain to the verge of bankruptcy in the early seventeenth century, by capturing Spanish ships laden with treasure from the New World, a desperate King Philip IV offered 10,000 ducats to anyone who could produce a workable SCUBA. Told by spies that as many as 500 ships at a time were harbored in the leading Dutch ports of Amsterdam and Rotterdam, Philip plotted to send divers into one or both ports during a storm and cut the ships' anchor cables, thereby causing collisions. But the Dutch had sentries on every ship, so the scheme would work only if the divers were totally concealed. A special committee was appointed to consider the designs that flooded in from every corner of the Spanish empire. None was deemed feasible, and the plan to disrupt Dutch shipping had to be abandoned.

Several of the designs that were entered in Philip's competition were interesting, though. Among them were three designs submitted by a Fleming, Florencio Valangren. Two of these weren't SCUBAs at all, while the third was a long breathing tube similar to those turned out regularly by inventors, which were useless beyond two or three feet below the surface. This one had an original touch: an exhaust valve at the bottom of the tube for exhaling foul air. The second also featured a tube; the end of the tube which was above the surface was connected to a bellows. This was the first known instance of air being pumped down to a diver, predating Dr. Papin by many years. The device could have worked in shallow water (the primitive bellows lacked the force to pump air to deep water). According to the letter accompanying Valangren's designs, similar devices were then in use in Flanders and Holland, primarily for repair work on ships' bottoms. The third design, a very primitive SCUBA, consisted of a large animal hide pumped full of air by a bellows and dropped by weights to the seafloor. Connected to the hide was a tube with a mouthpiece which could be opened when the diver, swimming freely around in the water, needed to take a breath. In his letter, Valangren stated that this device, like the tube-bellows contraption, was then in use. We have no reason to doubt him. With an air bag made from a large cowhide and lowered in about 20 feet of water, the diver would have been supplied with at least 15 minutes of air.

Another invention sent to Philip, this one anonymous, is even more interesting, for it went further toward developing a practical, self-contained underwater breathing apparatus. It consisted of a tubular air reservoir worn like a belt. Probably it was made of an animal's intestines. At one end there was a mouthpiece and at the other a bellows by means of which the diver could obtain air from the surface without assistance. The air reservoir was too small to hold much air, but if a diver was good at holding his breath and filled his lungs once a minute, he could go as long as 10 minutes before having to replenish the reservoir.

The next significant attempt to develop a workable SCUBA was by an Italian named Giovanni Borelli in 1679. His design was based on the principle of air regeneration: the diver using it supposedly could breathe the same air over and over again after his spent air was passed through a filter and freshened. The diver in Borelli's drawing is wearing a goatskin suit, swim fins, and a metal helmet with a window. The air reservoir is the helmet itself, and a copper tube (the filter system) leads from the front of the helmet to the back. Borelli claimed that condensation would collect in the tube; when the diver exhaled, this condensation would filter out the carbon dioxide from the spent air. Another invention of Borelli's was a cylindrical piston apparatus carried by the diver to help him descend and surface. To sink, he was supposed to open a valve and let in water, thus adding to his own weight; to rise, he was to let out the water, thus increasing his buoyancy (how he was supposed to get the water out is a mystery). Both the air-regeneration and piston-apparatus methods seem like the schemes of an armchair theorist—visionary designs that were far beyond the technological capabilities of the age. Probably neither got past the drawing-board stage.

In 1726 Stephen Hales, an English botanist, devised a unique apparatus which was based on the closed-circuit rebreathing principle used to combat noxious gases in unflooded mines. His apparatus consisted of a leather breathing bag, the walls of which were lined with flannel soaked in a mixture of sea salt and tartar (sodium chloride and phosphate of lime). A curved tube led from the breathing bag to a large wooden mouthpiece clamped between the wearer's teeth. It was thought that the salt and tartar mixture would absorb the "carbonic acid" of the exhaled air, which was rebreathed over a period of eight to ten minutes in a noxious atmosphere. Hale's device marked the emergence of the first workable closed-circuit rebreather in which the same small amount of air (later oxygen) was breathed over and over again, being cleared of its carbon dioxide content with a chemical absorbent. But its user courted death from anoxia when air and not pure oxygen was breathed. (The chemical element, oxygen, wasn't isolated until 1774.) Hale's apparatus was tested at Plymouth Harbor in 1731, but the test ended abruptly when one of the divers using it drowned.

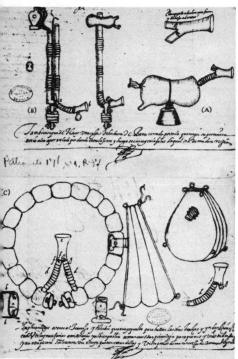

MARX

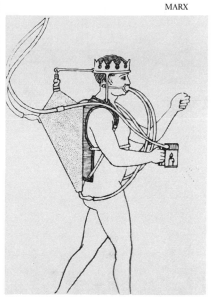

LEFT: *Three 1631 inventions by the Fleming Valangren: (A) a breathing tube with exhaust valve on the bottom; (B) a breathing tube with a bellows on the top; (C) SCUBA with air reservoir, bellows, and mouthpiece.*

RIGHT: *Drieberg's Triton, 1808. This diver carried a container holding a candle, and air was pumped in to keep the candle alight.*

MARX

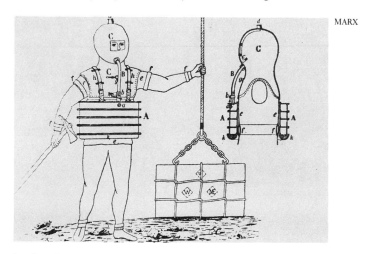

The first full-time SCUBA, invented by William James in 1825.

Two-man pigs were used by the Italians in attacks against Gibraltar during World War II.

View of damage done to H.M.S. Valiant *in December 1941 by Italian frogmen riding "pigs."*

Freminet, the Frenchman mentioned in Chapter Five, was the next to try his hand. He built a SCUBA by connecting an iron cylinder of air to the diver's helmet. After several unsuccessful trials, Freminet took a different tack and, as a result, made a significant contribution. On June 7, 1774, before witnesses, he stationed a man on the surface with a bellows, to pump a steady flow of air through a hose to the diver below. Hale dived to a depth of 50 feet and remained there for almost an hour. The success of this diving apparatus, coinciding as it did with work being done in England, pushed the SCUBA into the background for most of the next century. Working with the bellows concept, however, Freminet apparently attempted to develop the SCUBA further. There are numerous references to two such projects. In the first one, Freminet packed a bellows-in-a-box on the diver's back; and in the second, he sank a bellows-in-a-box alongside the diver, with a hose leading to the diver's mouth. Details of the projects aren't very clear, and there is no indication that either project was successful. Freminet's chief contribution to the development of the SCUBA was largely a negative one: he developed a successful non-SCUBA, then shifted his interest to air-hose diving.

In 1776 another Frenchman, a man named Sieur, actually built an air-regenerating system that was virtually a duplicate of Borelli's, and discovered that the filter would not purify the air. He admitted failure and then, undiscouraged, turned the invention into a helmet diving rig by attaching a hose to the helmet and having air pumped down by the bellows system Freminet had designed.

In 1808 a German, Friedrich von Drieberg, invented an apparatus which he called "the Triton." Resembling today's aqualung, it consisted of a metal cylinder filled with air, which was worn on the diver's back. A breathing tube led to the diver's mouthpiece. Most of the time, though, the Triton wasn't self-contained: tubes led from it to a pump on the surface. In the event of a pump failure or some other emergency, the diver had enough air to stay below for quite some time.

The first workable, full-time SCUBA was invented in 1825 by an Englishman named William James. A cylindrical belt that extended from the diver's waist to his armpits served as an air reservoir. The belt held air under a pressure of 30 atmospheres (about 450 pounds per square inch), and the diver breathed through a hose connected to this helmet. He wore a full suit, carried weights, and had boots instead of fins, which makes it clear that the rig was intended for walking on the sea bottom rather than swimming.

About this time, Charles Condert, a Brooklyn factory worker, also developed a workable SCUBA. Considerably different in appearance from James's, it consisted of a full diving suit and attached hood, both of heavy cloth overlaid with gum rubber. There was a tiny hole at the top of the hood for releasing spent air. The air reservoir—a four-foot length of copper tubing six inches in diameter and bent like a horseshoe around the diver's body—had a safety feature that James's gear lacked: a valve which allowed the diver to let air into his suit at will, thus keeping it from being forced against his body by the water pressure. Condert successfully used his diving suit at depths up to 20 feet for various underwater jobs around the Brooklyn docks. He died in August 1832 when the valve cock between his air reservoir and suit broke down as he was surfacing, and he became entangled in an underwater obstruction.

In the race to perfect a SCUBA, the French lost some prestige when in 1842, a Frenchman named Sandala announced that he had invented one. The apparatus turned out to be an exact copy of the one Sieur had copied from a design of Borelli's that was old. In spite of the fact that high-pressure air compressors already existed, Sandala designed a cylinder that held air at a pressure of only one atmosphere, which rendered it practically useless. Sandala was severely criticized by both the French and British press. It took the efforts of two other Frenchmen to restore France's prestige among the world's divers.

In 1865 Benoît Rouquayrol, a mining engineer, and Auguste Denayrouze, a navy lieutenant, designed a "self-contained diving suit." Actually, the suit wasn't completely self-contained. It depended on air being pumped from the surface to a metal cylinder on the diver's back, where a pressure of 40 atmospheres was maintained. As was true of Drieberg's "Triton," it could be disconnected and function for short periods of time on the air in the cylinder.

The most innovative feature of the Rouquayrol-Denayrouze diving suit was an attachment to the air reservoir which regulated the amount of air passing through the tube leading to the diver's mouthpiece. This regulator contained a membrane that was sensitive to the outside water pressure. As the pressure increased, the membrane would cause a valve to let air into the diver's suit at equal or greater pressure. The regulator sent air into the diver's mouthpiece only when he took a breath, rather than constantly, so that no precious air was wasted. Because the diver used up air pumped down to him fairly slowly, the surface pumps (in those days, limited in the amount of force they could exert) didn't have to be as efficient as pumps used to pump air to a helmet diver. Thus the device was more practical at greater depths than were the helmet diving rigs of the time. A very precise mechanism, the membrane regulator was the most important single device in the development of the modern SCUBA.

Originally the Rouquayrol-Denayrouze apparatus was designed to supply air to a nude diver wearing goggles, but the inventors soon recognized the danger of having the goggles flattened against the eyes, and replaced them with a face mask to which the hose leading to the cylinder was attached. The diving suit proved so satisfactory that many helmet divers began to wear air reservoirs on their backs, so they wouldn't have to depend on surface attendants for air. Commercial production of the suit began in 1867. Soon it was officially adopted by most of the world's navies. The diving suit was used more widely than any other type of diving equipment until the turn of the century when the air reservoir was transferred to the surface for safety reasons. If the compressor failed, a hand pump could force air into the reservoir. Thus the diver could continue to breathe until the emergency was over.

In 1867 T. S. McKeen had obtained U.S. patent number 65,760 for a "self-contained diving dress." There were actually two versions of this dress. The first consisted of a suit and helmet. Air was carried in a copper cylinder worn on the diver's back. The second was a kind of armored suit with spiral tubes containing air and encircling the diver from ankles to armpits. Presumably the air was compressed somehow, although how isn't clear. Nor is it known whether the units were ever built and tested.

In 1876 an English merchant seaman, Henry Fleuss, invented a simple, compact, lightweight, self-contained diving apparatus which used pure compressed oxygen rather than compressed air. It was based on the air-regeneration principle and actually worked. Although the device did not permit divers to reach even half the depth of the Rouquayrol-Denayrouze apparatus (pure oxygen is poisonous below 33 feet or when the person breathing it exerts great physical effort), it did make divers completely independent of a surface air supply, as the inventions of James and Condert had done. Its primary advantage lay in the three-hour air supply it carried.

Fleuss's apparatus (usually referred to as a "closed-circuit oxygen rebreather" rather than a SCUBA, even though it operated on a similar principle) closely resembled that of Rouquayrol and Denayrouze. The components were an air reservoir, regulator, and an air hose. The filter system was the heart of the invention. Caustic soda absorbed the carbon dioxide from the diver's breath, and he could breathe the remaining oxygen over and over again. The rebreather could be used whether the diver wore a suit or swam nude, whether he chose goggles and mouthpiece or a mask or even a helmet. Versatility was the thing.

The Fleuss rebreather proved its worth in 1870. A tunnel being constructed under the Severn River near Bristol, England, flooded. Before the tunnel could be pumped dry, a large iron door had to be closed. Divers wearing the standard helmet gear couldn't do the job. Twists, turns, and obstacles in the tunnel made it a near certainty that their air hoses would snag or be cut. The Rouquayrol-Denayrouze apparatus wouldn't work either. Disconnected from the surface, it wouldn't have supplied enough air for the duration of the work. Fleuss hired the reknowned helmet diver, Alexander Lambert and taught him to use the rebreather. Lambert succeeded in closing the door, but three years later, when the tunnel flooded again, he nearly lost his life from oxygen poisoning which resulted from his extreme physical exertion.

Because of the danger of oxygen poisoning, the Fleuss rebreather was limited to operations that did not tax a diver's energy unduly. Between 1900 and 1910 it was adapted as a submarine escape device even though it was hazardous to use below 50 feet. A diver could safely ascend from a depth of several hundred feet, provided he was quick about it. The Fleuss rebreather had an advantage over other diving equipment, which made it invaluable in military operations: total concealment for the diver. During World War II it was preferred over all other SCUBA gear, even the aqualung which was a later and more practical device for most purposes but which left a telltale stream of bubbles in its wake, much as the early snorkels had done.

In 1900 a French professor of biology at the University of Algiers, Louis Boutan, finding no diving equipment suitable for underwater photography, devised his own SCUBA. Along with the standard helmet diving suit that he wore on his back, Boutan installed a large cylinder of compressed air capable of sustaining him for up to three hours at a depth of 70 feet. As far as we know, no one else used Boutan's SCUBA. Apparently he never seriously tried to promote it.

Just before the turn of the century, Georges Jaubert discovered "oxylite," a kind of oxygen in a latent state. The powder (potassium and sodium peroxide) changes into pure oxygen gas when it comes in contact with a small amount of water. Too much water, and the powder reacts violently, producing a lot of heat and smoke—and, in addition to oxygen, an extremely toxic solution. The potential danger of leakage in using it underwater is obvious. Nevertheless, in 1904, the London firm of Siebe, Gorman and Company acquired the patent rights to oxylite and incorporated it in a submarine-escape device. This was done under the direction of Sir Robert Davis, the director of the company and a pioneer in the development of diving equipment. Although oxylite has been replaced by a cylinder of oxygen, the Davis Submarine Escape Rebreather is still used today as an escape device aboard submarines.

Then the Germans got into the act. In 1905, Draegerwerk, a company which manufactured helmet diving equipment, introduced its version of the Davis Rebreather. Originally intended for mining, it was soon converted for underwater use. The sinking of the French submarine *Pluvoise* in 1910 was the occasion of a sudden renewed interest in rebreathers, as well as all types of SCUBA. There was no escape device on the sub; in fact, there were none in the French navy at the time. Draegerwerk introduced a new submarine escape apparatus which consisted of a breathing bag, potash regenerator, a supply of compressed oxygen, and a hand-operated intake valve. When using the unit, the diver also wore a nose clip.

In 1912 military necessity was the reason for another SCUBA development. The new development was a self-contained submarine sled which was ridden by a diver. The sled was the invention of a German submarine officer, Captain Max Valentiner, and was produced by Draegerwerk. The mission of the sled, which was towed by a surface craft, was to enable a diver scanning the bottom to find torpedoes lost during submarine practice firing, as well as for other military uses. The SCUBA unit included with the sled was described as follows: "The regenerating apparatus comprises steel cylinders filled with compressed oxygen, a potash cartridge and a circulating arrangement, viz., a suction and pressure reduction valve. It is connected with the helmet by two short lengths of hose, one of which serves to draw off any used-up air, while the other supplies fresh air." The unit was described as adequate for two to three hours underwater work, "in accordance with the personal skill of the diver and with the quantity of carbonic acid secreted by his lungs." It was declared successful to a depth of 66 feet. But even then, work was underway on other SCUBA devices which would allow deeper descents. Early models were equipped with two large tanks of compressed air to allow the diver to sink or rise as desired; in later models, the tanks were discarded, since the sled was being towed and they were unnecessary.

Also in 1912, the Westfalia Maschinenfabrik of Gelsenkirchen, Germany introduced a self-contained diving dress in which a mixture of 45 percent oxygen and 55 percent nitrogen was used for descents to 100 feet. In another apparatus manufactured by the company, a mixture of 30 percent oxygen and 70 percent nitrogen was used for descents to 200 feet. (Normal air is 79 percent nitrogen and 21 percent oxygen.) This was probably the first use in SCUBA of special mixtures of oxygen and inert gas to reduce the toxic effects of the compounds. For some reason, though, neither unit was ever widely distributed. The following year, Draegerwerk produced a similar unit, which used a mixture that was 60 percent oxygen and 40 percent nitrogen and which was mixed automatically. A few years later, Siebe, Gorman and Company introduced a SCUBA which consisted of two steel cylinders containing 10 cubic feet of air and oxygen each. The contents of the cylinders were compressed to 120 atmospheres and mixed as drawn upon by the divers. This SCUBA was useful only down to 66 feet; beyond that, the diver risked blacking out. The Germans went one step better in 1914 when they produced a SCUBA that used three cylinders which were worn on the diver's back and contained only air. Then World War I broke out, and SCUBA development practically came to a halt.

In the years before the war, a French navy captain, Yves le Prieur, became interested in diving with the traditional helmet-hose, surface-supplied-air equipment and in swimming underwater without any equipment. He soon found that this approach had both advantages and drawbacks. Helmet diving allowed reasonably long periods of time underwater but limited freedom of movement laterally, whereas skin diving, with no equipment except a pair of goggles, allowed great underwater mobility but extremely limited bottom time. Le Prieur hoped to combine the best of both methods. In 1926 he and a countryman named Fernez obtained a patent for the "Fernez-le Prieur self-contained diving apparatus." Their unit included a steel cylinder with a capacity of three liters, into which compressed air up to 1,950 pounds per square inch was pumped. The cylinder was worn on the back, with a short air hose leading to a mouthpiece held in the diver's mouth. A pressure gauge jutted over the left shoulder. The diver wore small tight goggles and a nose clip. The equipment weighed about 20 pounds and reputedly allowed about 10 minutes underwater at depths of less than 50 feet.

In 1933, when the sport of underwater spear fishing, or skin diving (so named because it resembled the birthday-suit free diving that had been practiced for centuries), was becoming popular in the Mediterranean, le Prieur received a patent and began manufacturing a commercial SCUBA for the growing band of enthusiasts. His unit consisted of a steel cylinder of compressed air worn on the diver's back and connected by an air hose to a full face mask. There was no regulator, just a valve with which the diver controlled the amount of air flowing into his face mask. Without a regulator, however, air pressure equivalent to the surrounding water pressure couldn't be maintained, which meant that the unit could be used safely only in shallow waters. There was also a constant flow of air into the diver's mask, and, since he couldn't breathe it all, much of it went to waste. Still, the cylinder did allow the diver to spend 30 minutes at 23 feet, 20 minutes at 33 feet, or 10 minutes at 40 feet.

Le Prieur's SCUBA was adopted by the French navy in 1935, and the next year he founded the first SCUBA diving club in the world, which he called the "Club of Divers and Underwater Life." Its members staged a human aquarium at the International Exposition in Paris in 1937, to demonstrate their abilities underwater. That same year, another Frenchman, Georges Comheines, revised le Prieur's SCUBA. Comheines used two or three cylinders of compressed air, each with of capacity of four liters. The three-cylinder apparatus allowed up to 40 minutes at a depth of 33 feet or 25 minutes at 66 feet. Although Comheines was drowned during one of his first dives, his equipment was widely acclaimed and used.

One of the most puzzling aspects of the history of the SCUBA is the long neglect of the Rouquayrol-Denayrouze regulator. Even though the rest of their apparatus had been made obsolete by new equipment, the regulator was a success. Why was it ignored? Neither Boutan nor le Prieur used it. Not until 1942 was it brought back into use, and then by Cousteau and Emile Gagnan, an engineer. Cousteau was a pioneer of skin diving in the Mediterranean. He wanted to reach greater depths than the le Prieur unit would permit. He and Gagnan combined a regulator similar to the one in the Rouquayrol-Denayrouze apparatus, with le Prieur's compressed-air cylinder. They made several minor improvements in the air hose and mouthpiece and created the celebrated aqualung.

With the aqualung, skin divers have reached depths never before possible, where they remained for longer periods of time, the length depending on whether they carried one, two, or three cylinders of air. In October 1943 a close friend of Cousteau's, Frédéric Dumas, made a record dive to 220 feet, and in August 1947 he reached 307 feet. A few weeks later, Maurice Fargues, a French naval officer, reached 397 feet but got nitrogen narcosis. Somehow he managed to reach the surface where he lost consciousness and died, probably as a result of contracting the bends by surfacing too fast. Because of other deaths from nitrogen narcosis, it's considered unsafe to descend below 200 feet on compressed air. If helium-oxygen mixtures are used and diving tables are followed, however, a diver with an aqualung can go as deep as a helmet diver can.

DIVERS IN WARFARE

Chronicles of divers taking part in warfare date back to the beginning of recorded history. We have seen that they were active in battles fought in the warm waters of the Mediterranean, but they were also active in the frigid waters around Greenland, Iceland, and Scandinavia. There are many medieval accounts of their use by the Vikings, as well as their enemies, for boring holes in ships' bottoms, cutting anchor cables, and committing other acts of sabotage. During the Crusades they were often used by both the Christians and the Saracens for underwater sabotage and reconnaissance. Breathing through snorkels and sometimes wearing goggles, diving scouts entered enemy ports to learn troop strengths, the size of fleets, and the layout of fortifications. Since there were few reliable charts before the sixteenth century, another important task they were entrusted with was sounding the entrances to harbors so their own ships wouldn't run aground or onto rocks. Over the centuries, methods of sabotage and reconnaissance changed little; military strategists were slow to extend the range of wartime activities assigned to divers.

A man who grasped the military potential of divers quite early was an Englishman, the Marquis of Worcester, who informed his king in 1748 that he had invented "an engine very compact and portable, small enough to fit in one's pocket, which could be used to sink a ship no matter what its size, and after a diver had fastened this engine to the under part of a ship, it could be exploded at any appointed minute, tho' a week later, either by day or night, and totally destroy said ship." The marquis' device was a forerunner of the limpet mines used in World War II by the Italians and British with great success. But in the eighteenth century it was ignored, partly because advisers to the British Admiralty insisted that the claims for the mine were exaggerated and partly because the marquis demanded an exorbitant sum of money before he would hand it over to the king. Although explosives were sometimes attached to enemy ships by submarines (David Bushnell, inventor of the submarine boat, the predecessor of the modern submarine, devised a way to do that), divers did not handle bombs until the present century.

A modern invention similar to the one proposed by the Marquis of Worcester was described by the American inventor, Simon Lake. Lake, who collaborated with John Holland in designing several submarines, wrote an article for *Scientific American* in 1915, outlining a plan for building special trapdoors on submarines, so divers could leave to plant explosive charges with delayed action fuses on the bottoms of enemy ships and return to the subs before the explosives went off. The divers would be helmet divers equipped with air hoses connected to the submarines. This implied that the submarines would have to be very close to their targets or, if the divers had long hoses, risk snagging the air lines on underwater obstructions. The U.S. Navy rejected Lake's plan, arguing that even if a submarine did get close enough to an enemy vessel to make the use of helmet divers practical, it was still simpler to use torpedoes. True enough, but perhaps Lake's proposal would have been taken more seriously if he had suggested furnishing the divers with oxygen rebreathers.

About this time, the British Admiralty, then headed by Winston Churchill, was rejecting another plan to extend the use of divers in naval sabotage. The plan, which was the work of a British naval officer, called for the use of torpedoes which, instead of being fired from submarines, would be controlled by divers wearing oxygen rebreathers and riding on the torpedoes. A diver would steer for his target. When he reached it, he would attach the torpedoes to the ship's hull with magnets, activate the delayed-action fuses, and leave. Not surprisingly, the plan was called the "human torpedo." Churchill rejected it on the grounds that it would not be humane to risk the lives of divers in such an operation, for he thought it highly unlikely that they would escape safely.

A country which used divers to handle munitions well ahead of the rest of the world was Italy. Even before World War I the Italians had a school for training divers in underwater sabotage, with the emphasis on techniques of fastening explosives to enemy vessels and getting away safely. When the war began, the Italians came up with the "chariot." A small craft 33 feet long and powered by compressed-air engines, it was designed to run submerged two or three feet below the surface. The chariot carried two powerful mines, each capable of sinking a large ship. The diver, who wore no equipment except goggles (the Italians considered oxygen rebreathers too dangerous for the job) rode the chariot with only his head above the water. When he neared his objective, he would detach the mines, leave the chariot, and swim underwater to the enemy ship. After attaching the mines to the hull with magnets and setting the delayed-action fuses, he would escape in the chariot.

The first chance to use the chariot came in 1918. The Allies decided that, whatever the cost, some damage had to be inflicted on the Austrian fleet, which had been bottled up in the Adriatic port of Pola since 1914. For four years defenses such as antisubmarine nets (steel nets which were suspended from cables floating on the surface, to keep out subs) had made the fleet invulnerable. Ever since a French submarine had been caught in one of the nets and destroyed, no other submarine had attempted to enter the port.

The Italians chose their two best divers, Major Raffaele Rossetti and Lieutenant Raffaele Paolucci, to command the two-man chariot. On an overcast night, a patrol boat dropped them a quarter of a mile outside the Pola breakwater. At the breakwater entrance, the divers struggled against an outgoing tide to pull the heavy chariot over the antisubmarine net blocking their way. Finally they succeeded. But this was only the first of several obstacles. Soon Rossetti and Paolucci reached the next net and farther on, still another. This one was topped by a boom bearing three-foot iron spikes. As boats carrying sentries rowed past, the divers pulled the chariot over the boom. Fortunately, at this point, it began to rain, and they surmounted three more antisubmarine nets under its cover. Several times they found themselves so close to the sentries that they could understand what they were saying. At last they reached the warships and chose the largest, the *Viribus Unitis*, as their target. It took them two hours to attach the mines to the ship's keel and set the fuses for 6:30.

It was already 5:15. The divers climbed back into the chariot, only to find that most of the compressed air for the motor had been consumed in their struggle to reach the Austrian warships. The amount left wasn't enough to get them back to the patrol boat, so they decided to sink the chariot and swim for shore. But they were soon sighted, captured, and brought aboard the *Viribus Unitis*—the very ship they had fastened the mines to. Rossetti and Paolucci thought they were doomed. But another kind of shock awaited them. They discovered that their captors were Yugoslav, not Austrian. The Yugoslavs were Italy's allies. While the patrol boat was taking them and their chariot to the harbor at Pola, the Austro-Hungarian empire had collapsed, and the fleet anchored in Pola was in Yugoslav hands now.

Rossetti and Paolucci immediately demanded to see the captain. They quickly explained their mission and warned him that his ship would be blown up at 6:30, which was now only a few minutes away. The captain didn't believe them, and, because they were foreigners who had been caught where they had no business being, he sent them to the ship's brig. As the minutes ticked away, they pleaded with their guards, who mocked them, laughing hysterically. The mockery increased as 6:30 came and went and nothing happened. Then, almost an hour late, two explosions rocked the ship. Water began to pour in through two enormous holes in the hull. The captain gave the order to abandon ship and ordered that the prisoners be released from the brig. Several hundred Yugoslavs died, though. Rossetti and Paolucci survived and were allowed to return to Italy when it was was proved that they had made an unintentional, if deadly, mistake. At home they were awarded 650,000 lire by the Italian government. Overcome with remorse for the holocaust they had caused, Rossetti and Paolucci donated the money to the widows of the dead Yugoslav sailors.

The chariot had been proved a success but at an extravagant cost. The incident should have been a lesson to the major world powers about the military effectiveness of divers. But none paid any heed. No country took steps to train its divers for underwater sabotage, perhaps because it was widely believed that World War I was the war to end all wars. Meanwhile, the Italians increased their diver training program, improving their oxygen rebreathers, until by 1920 all their divers were adept at using them. The fruits of the efforts were eventually to be turned against their former allies.

In December 1934, Italy attacked Ethiopia. Fearing that Great Britain would intervene on behalf of the Ethiopians, Italian military leaders sought a means of crippling the powerful British fleet in the Mediterranean. Two submarine officers, Lieutenant Elias Toschi and Lieutenant Teseo Tesei, proposed a plan calling for human torpedo riders—virtually the same plan Churchill had rejected 20 years earlier. The Italian military accepted it. Toschi and Tesei designed and built several torpedoes, which they nicknamed "pigs," and trained divers in the Italian navy to use them. But Britain did not intervene, so the project was shelved for the time being.

When, in 1939, the major powers were preparing for war, the Italian high command ordered Toschi and Tesei to resume work on their torpedoes. This time, they designed larger and more powerful pigs. Two divers were to ride on each pig, which were 22 feet long. The machine could make three knots; it had a range of 10 miles and could reach a depth of 100 feet. In spite of the danger in using oxygen rebreathers at that depth, the divers had to wear them. Otherwise, the stream of bubbles would make them too conspicuous. The plan called for a surprise attack that would immobilize the British Mediterranean fleet as soon as the war began.

Before the redesigned pigs had been fully tested, however, war was declared. The Italians decided to use them anyway. Late one night in August 1940, Toschi, commanding a unit of volunteers known as the Tenth Light Flotilla, led an attack against the British fleet, then anchored in the harbor at Alexandria. The raid failed. Just as the submarine carrying the pigs and divers surfaced to release them, British airplanes spotted it and bombed it. Toschi and most of the submarine crew escaped, but were captured. The next month, the Italians planned two more attacks using divers and pigs. One attack would be on Alexandria and the other on Gibraltar. Both attacks had to be cancelled, however, when it was learned that the warships they expected to find in the harbors had put to sea. In October another attack was attempted at Gibraltar, but it failed when the four pigs that were being used broke down and couldn't reach their objectives. Probably salt water had seeped in, shorting out their batteries. The pigs had to be scuttled. Six of the eight divers on the mission swam to the Spanish side of the bay and made their way back to Italy. The other two were captured but never told the British of the details of their plan.

In May 1941 still another attack against Gibraltar was launched. This one was also unsuccessful. Two pigs developed mechanical trouble, and on the third, the breathing apparatus of one of the divers riding it failed. A July attack launched in Malta also came to nothing again because of mechanical failure. On the seventh try, made that September, the Italians at last succeeded. Three pigs launched from a submarine off Gibraltar penetrated the antisubmarine defenses of the harbor, and three British warships totaling 30,000 tons were sunk. The British were mystified; the attack had seemed to come from nowhere. No Italian diver in captivity had talked, nor had any of the pigs that had failed during earlier attacks been found (they were scuttled); British intelligence began an investigation.

The Italians were to follow their attack at Gibraltar with an even more impressive raid. In the fall of 1941, not long before the United States entered the war, British naval resources were strained to the limit. Many of the ships in their Mediterranean fleet had been transferred to the north Atlantic to help guard against the threat to Allied shipping by the U-boats. There were also nighttime raids in the North Sea by the German battle cruiser *Tirpitz,* then the world's most powerful warship. In early December, as the British in North Africa were falling back before Rommel's forces, U-boats sank the British aircraft carrier *Ark Royal* and the battleship *Barham.* These losses left Britain with only two large battleships, *Queen Elizabeth* and *Valiant,* and a few smaller ships to confront the Italians in the Mediterranean. When Italian aircraft spotted the two battleships off Alexandria, the Italian high command ordered them destroyed or immobilized at all costs.

The Tenth Light Flotilla prepared to attack. Only three "pigs" were available, however. Six divers boarded the submarine *Scire* which carried them to Alexandria. Carefully threading her way through the mine fields around Alexandria, the *Scire* surfaced during the night a few miles outside the harbor, and the divers (for the first time wearing the swim-fins that prompted the British to call them "frogmen") and their pigs were launched. Because it was too dangerous for an Italian submarine to remain in these waters, the *Scire* headed back as soon as it had discharged its passengers. After completing their job, the frogmen were to swim ashore, there to make contact with the Italian underground who would smuggle them out of Egypt and back to Italy.

The frogman ran the pigs on the surface at full speed until they were just outside the entrance to the harbor, where they were in danger of being spotted. Instead of going over the six antisubmarine nets protecting the anchorage, they decided to try to go under them. This would be extremely dangerous; they would have to dive much deeper than was safe with the oxygen rebreathers. Somehow they made it, though, only to find themselves up against a new hurdle. A few days earlier, the British had learned of the human torpedoes. Now they were dropping depth charges every few minutes, to either kill the divers or force them to abandon their attack. Not dissuaded by the periodic blasts, which shook them and punctured their eardrums, the frogmen struggled on courageously. Each pig reached its objective: the *Queen Elizabeth,* the *Valiant,* and a large oil tanker, the last chosen in hopes that the explosion would ignite other ships in the harbor. And explode, the tanker did, bursting into flames and igniting several other ships. Large holes were also blown in the two battleships which settled to the bottom, with only their superstructures sticking above the surface.

It took the British months to repair and salvage the *Queen Elizabeth* and *Valiant,* months when they were seriously understrength in the Mediterranean. Fortunately for them, but not the Italians, all six frogmen were captured along with the members of the underground. Thus Italy didn't learn for some time of their success. Churchill later said that if the news had been known at once, the war might have taken a different turn; for the Italian navy would have committed itself, and this could have wreaked great havoc throughout the Mediterranean.

This wartime incident was by no means the last achievement of the Tenth Light Flotilla. By the time Italy surrendered in 1943, its frogmen had accounted for the sinking of more than 30 warships, as well as damaging many others. In fact, their exploits made those of the rest of the Italian navy look like small potatoes.

The frogmen enjoyed another success, one they hadn't anticipated—converting the British to staunch believers in the effectiveness of frogmen in warfare. Seldom backward when it came to maritime affairs, the British Admiralty began a crash program to train frogmen. They also adopted the use of pigs and developed other devices such as small submarines to carry frogmen into enemy harbors where they could attach limpet mines to the ships, much as the Marquis of Worcester had proposed two centuries earlier. Like the Italians, the British frogmen failed many times before scoring a success. During an attack on the *Tirpitz* in October 1942, four pigs that were being towed to the ship's anchorage were separated from their towlines because of rough weather. Two other attempts to use pigs against the *Tirpitz* failed, due to mechanical failure. During the Allied landing at Palermo in January 1943, the British finally succeeded. Sixteen frogmen on eight pigs sank six Italian ships—a cruiser, three submarine chasers, and two merchantmen. The Italians had gotten a dose of their own medicine.

Owing to the strict military secrecy Great Britain maintained until well after the war, it wasn't until 1948 that the world learned of the greatest triumph credited to British frogmen. On September 22, 1943 another attack was made against the *Tirpitz* in Käfjord, Norway. Because of the shortage of rubber needed for suits to protect divers from the frigid water, the British decided to use their newly designed midget submarines, or X-craft, which carried four-man crews and had large explosive charges and were launched underwater from a standard submarine. Six were to take part in the operation. However, the submarine carrying three of them was lost en route, and, of the remaining three, one was never seen again after being launched at the entrance to the fjord.

The remaining two X-craft penetrated the antisubmarine defenses and dropped their charges underneath the *Tirpitz* without their frogmen having to go out in the icy water. One X-craft got away from the *Tirpitz* but ran afoul of an antisubmarine net and couldn't get loose. Since there were only two oxygen rebreathers on board, two members of the crew died. The others were caught and taken aboard the *Tirpitz* where they joined the crew of the other X-craft, which had also been captured. While the Germans interrogated their prisoners, a tremendous explosion ripped through the ship. To keep her from sinking, the *Tirpitz* was quickly run ashore. Some eight months went by before she was back in action, which gave the British a badly needed breathing spell.

Striving to improve their midget submarines, the British developed a more advanced model called the XE-craft, with increased speed and a greater range. In the new model, explosive charges were to be attached to the bottom of an enemy ship by a frogman rather than dropped under the hull. To leave the XE-craft, a frogman entered the escape chamber and closed the door. The chamber was then flooded and a valve was opened to release the trapped air. As soon as the pressure inside was the same as the outside water pressure, the frogman opened the hatch and left. On his return, the procedure was reversed.

The XE-craft was first used in July 1945 in an attack on two Japanese heavy cruisers, the *Nachi* and the *Takao*. Submarines towed the two midget craft from Borneo to within 40 miles of the cruisers' anchorage in the Johore Straits near Singapore. One of the XE-craft had to turn back because of engine trouble, but the other one continued on alone, running on the surface by night and submerged by day. It took the XE3 a day and a half to make the trip.

The plan called for a night attack, but by the time the XE3 reached the *Nachi* and the *Takao*, its batteries were so low that the commander, Lieutenant Ian Fraser, had to attack at once. He chose the nearer cruiser, the *Takao*, as his target. Dropping from periscope depth when he was less than a mile away, he ran at a depth of 30 feet straight for the cruiser. Suddenly Fraser noticed that the water was becoming very shallow. Before he could do anything about it, however, his XE3 was wedged between the *Takao* and the bottom. But the Japanese hadn't heard the collision nor spotted the XE3, although it was noon and the water was very clear. Believing that they would be spotted at any moment, Fraser ordered Seaman McGinnes to attach the limpet mines with a very short fuse, so they would explode before the Japanese could send divers down to remove them. McGinnes couldn't open the exit hatch, though—they were wedged against the keel of the *Takao*. For nearly an hour they did everything they could think of to move the XE3 enough to open the hatch part way. Finally, McGinnes, his body coated with grease, managed to inch his way out, but he tore the bag of his oxygen rebreather in doing so.

He took six mines from the munitions locker outside and went to work, only to find that the bottom of the *Takao* was covered with a thick growth of barnacles, which prevented the magnets from clinging to the hull. Tension mounted inside the XE3. A task that should have taken only a few minutes dragged on interminably. Meanwhile, McGinnes scraped laboriously to get six areas clean. By the time he had finished, he was dangerously weak from breathing a mixture of salt water and oxygen. But he managed to crawl back through the hatch before passing out. Fraser immediately started the engines, but the craft wouldn't budge. He tried backing away, blowing his ballast tanks—anything he could think of. Nothing worked, and time was passing. At last, just a few minutes before the charges were due to go off, they broke free and got out of there. As the XE3 passed under the first antisubmarine net, Fraser and his men heard the explosion. The *Takao* sank almost immediately. Fraser and McGinnes were later awarded the Victoria Cross, Britain's highest military award.

The Japanese were slow to grasp the importance of divers in warfare, and their military plans for divers came too late to do much good. When the Americans occupied Japan after the war, they found 600 midget submarines on the production lines, but none were ready for action. The Japanese had developed a scheme to use divers, called *fukuryi*, much as the kamikaze suicide pilots were used. In the event of an American invasion of the mainland, several thousand *fukuryi* frogmen wearing oxygen rebreathers and rubber suits to protect them from the cold, and carrying long poles with contact mines at the top, were to be stationed underwater. They would jab at landing craft passing overhead, blowing up the invaders and themselves. There was no invasion, of course, and the desperate plan was never put into operation.

Divers played no more important a role in Germany's military strategy than they did in Japan's, although the Germans instituted a crash diver-training program similar to that of Great Britain's. Except for a few routine jobs in Holland, however, divers were used very little. Perhaps the Germans had too much confidence in their U-boats.

For some time the United States also relegated divers to obscurity. Despite the feats of Italian and British frogmen, the U.S. was slow to form its own corps of frogmen. Also, compared to British and Italian undersea technology, American methods were primitive.

It wasn't until the near disaster of Tarawa in November 1943 that the U.S. Navy recognized the need, and even then, the use of frogmen was limited. Tarawa was the first of a series of islands and atolls in the Pacific that the Americans had to capture before they could get at the Japanese mainland. The plan for taking Tarawa called for several weeks of intense bombardment, followed by landing a Marine division. The entire operation was expected to take only a few hours. Because the military planners had no charts of Tarawa, they had to rely on aerial photographs, which showed the reefs surrounding the atoll but gave no indication of how deep they were. It was thought that the water was deep enough for the landing craft to pass over the reefs—but it wasn't. When the attack started, the landing craft were stranded on the reefs, where they were sitting ducks for the Japanese artillery ashore. In addition, the size of the Japanese force defending the island was larger than had been estimated. The Marines aboard the landing craft had to disembark and wade the rest of the way virtually point-blank into the machine guns of the Japanese. Only a few reached the beach alive. More and more Marines were sent in, only to be mowed down, too. Finally, the ships laid down a smokescreen and a beachhead was established. In the three days of fighting that it took to capture Tarawa, more than a thousand men were killed and three times that number wounded. It was one of the bloodiest battles the Marines ever fought.

The mistakes at Tarawa were carefully studied. Military leaders saw that if divers had been available to sound and map the reefs, another strategy would have been used, thus averting much bloodshed. Orders were quickly dispatched to all units of the navy, requesting volunteers for extremely hazardous duty. Units of 100 men each, called "underwater demolition teams" (UDT) were formed and sent to bases in Hawaii, California, and Florida where they were given training in long-distance swimming and practice in the handling of explosives. There was no training in deep-sea diving; the navy couldn't provide breathing apparatus, swim fins, or face masks at the time. The first UDT men went into action equipped only with goggles and knives—equipment no more advanced than that used by Mediterranean divers during the Middle Ages.

Before the first group of volunteers had completed their training, about 20 of the best were taken from Hawaii and rushed to the invasion fleet on its way to Kwajalein Atoll. The navy had a secret weapon called the "Stingray," which was a fast boat loaded with several tons of explosives. The UDT men were to ride in them, getting as close as possible to enemy-made underwater obstacles or places in the reefs where a passage had to be cleared by blasting. Then they were to get aboard other boats and return to their ships while the Stingrays were guided to their targets by radio control. Of the dozens of Stingrays launched the day before the scheduled landing on Kwajalein, however, not one reached its objective. All of them exploded prematurely or ran off course and exploded in the wrong places.

The invasion was delayed, and an alternate plan—calling for the UDT men to swim along the reefs and find openings, was implemented. Fast boats dropped the UDT divers near the reefs, while warships lying offshore covered them with protective fire. They accomplished their mission, locating many openings and marking them with large buoys. Admiral Turner, the commander of the invasion force, was so pleased with their success that he ordered them used the next day to swim in over the reefs as close to shore as possible, in an attempt to locate enemy gun emplacements. In this, too, the UDT men succeeded. All the enemy gun emplacements were knocked out by shelling from the battleships before the troops were landed. After the invasion of Tarawa, frogmen played an important role in the American military strategy in the Pacific. Their methods quickly became less primitive as swim fins and face masks were hurriedly manufactured and rushed to them.

U.S. Army commanders in Europe realized that they too could use frogmen in the impending invasion of occupied France, and training school for navy frogmen was hastily set up at Fort Pierce, Virginia. But when the army saw that the navy wasn't sending enough men, it asked for volunteers and had them trained in a school set up in England. Although the Americans raced to get ready for the invasion of Normandy, the British were far ahead of them, in numbers, training, and equipment, which included oxygen rebreathers and rubber suits. However, both American and British frogmen prepared for their biggest challenge. The Germans had fortified every landing place. Beach obstacles ranged from contact mines atop poles stuck in the sand to elaborate steel hedgehogs shaped like children's jacks and filled with contact mines. The initial plan—calling for the landing craft to get as close as they could to the sand dunes fringing the beaches at high tide, when the obstacles were submerged—had to be abandoned. There weren't enough frogmen in the combined British and American forces to clear all the mines from the beaches selected for the landing. To clear the two main landing beaches for the American forces, Omaha and Utah, it was estimated that 1,000 frogmen would be needed. There were only 175 UDT men and 200 hastily trained U.S. Army frogmen. The plan was changed. The invasion was now set to go off at low tide when the mines on the beach would be exposed, and the frogmen would have a better chance of destroying them all.

To keep the Germans from being tipped off as to the landing site, the frogmen did not begin their task the customary few days ahead of time, but rather a few hours before the first waves of infantry went in. Thus speed was essential. Rubber rafts carried frogmen and explosives ashore, where, under cover of a smoke screen, they exploded thousands of mines. The lives of many infantrymen were no doubt saved. But some 50 percent of the American frogmen and 60 percent of the British were either killed or wounded.

Eight days after the Normandy invasion the largest American invasion in the Pacific until then got underway against the Japanese-held island of Saipan. Saipan was the gateway to Japan's inner defenses. Two hundred UDT men were sent out on reconnaissance for several days beforehand. The invasion went slowly, though; the defenders held out for weeks. Finally, the U.S. military leaders concluded that the only way to take Saipan was to first take two nearby islands, Tinian and Guam, which were supplying the Japanese on Saipan with reinforcements and supplies.

Five underwater demolition teams were used, two at Tinian and three at Guam. At Tinian, frogmen were sent to determine whether the two beaches chosen for the invasion were free of underwater obstacles. Because the frogmen had to work by moonlight, they painted their bodies silver. Then they crawled up on the sand to search for gun emplacements and camouflaged pillboxes. Their first night's mission revealed that both beaches were so heavily fortified (many of the guns were hidden in caves) that neither naval fire nor aerial bombardment would do much good. The admiral commanding the operation was ready to cancel the landing, but the Marine commander, "Howling-Mad" Smith, disagreed. He persuaded the navy to have the frogmen search for alternate landing sites, particularly small beaches which the Japanese, not anticipating landings on them, might not have fortified as heavily.

During the next two nights the two UDT units on Tinian were busy taking soundings of areas off every beach and marking the locations and readings on slates. Others penetrated the beaches and moved inland to locate gun emplacements, with knives their only weapons. At the end of each night's work, the information they gathered was sent to the flagship of the fleet and analyzed. Eventually, two landing sites were picked, and the island was taken with few casualties.

Meanwhile, the three teams of frogmen on Guam had their hands full. Unlike Tinian, there were only two suitable landing beaches. The water off the others was too shallow for landing craft. Well aware of this, the Japanese had spent four years building and installing underwater obstacles. In front of the beach selected by the American leaders for the main assault, they had built a two-mile-long fence of cribs made out of coconut-tree logs and filled with coral rocks. The cribs were connected by steel cables and barbed wire strung with contact mines—certain destruction for any landing craft that tried to get past. The fence stood between the beach and a barrier reef, itself a dangerous hurdle.

It was important that Guam be taken as soon as possible. The frogmen were given seven days in which to open up passages in the reef and fence. Even for 300 frogmen, this was a formidable job. They had to work by night and day, constantly exposed to rifle and machine-gun fire from the shore. Four days and nights were devoted to mapping the reef, selecting natural openings in it, and making new ones with explosives. The next three were spent blowing up large sections of the fence. The night before the invasion, frogmen were sent onto the beach to locate major gun emplacements so they could be knocked out before the troops were landed. On this last, dangerous mission, several cocky frogmen swam ashore with a large piece of wood which they planted in the sand as sniper bullets ricocheted around them. The next morning the first wave of Marines found a sign that read: WELCOME MARINES. COURTESY UDT-4. An arrogant gesture, but no one resented it. The frogmen had saved thousands of American lives.

Invasion followed invasion in the Pacific. Always the frogmen were used successfully, risking their lives perhaps more than any other men in the armed forces. There were few casualties, however, mainly because the Japanese could fire on the frogmen only with small weapons. If they used anything larger, they risked revealing the positions of their larger weapons to American ships and aircraft. Another reason was that the frogmen usually worked under cover of darkness. When they had to work during daylight, the navy provided them with smoke screens. Also, they stayed underwater as much as possible and avoided making themselves good targets for Japanese small-arms fire.

At Iwo Jima, the bloodiest battle in the Pacific during World War II, the UDT frogmen suffered their greatest losses, although they weren't actually on a mission. Four teams were to scout the landing beaches for mines and other obstacles, of which there were few, and collect samples of sand so the experts could determine whether it was firm enough to permit heavy tanks to be landed. Shortly after sunrise on February 17, 1945, 12 heavily armed landing craft, carrying 400 frogmen, and a small flotilla of small ships to lay down a covering fire, broke from the large American fleet and began to move toward the beaches. The Japanese, mistaking the mission for the actual invasion, opened up with everything they had. Within minutes, all of the landing craft had been hit. But most of the frogmen managed to reach the shore where they completed their mission. They suffered only two casualties. After one team had returned to their home ship, however, the ship took a direct hit from a Japanese bomber. Twenty-eight frogmen were killed and 43 wounded.

By the time of the invasion of Okinawa on April 1, 1945 the underwater demolition teams had come a long way. The navy now had 10 teams totaling 1,000 men for use in the invasion. Again, the frogmen performed missions similar to those of previous invasions. Even though they were used daily for several weeks, there were only about a dozen casualties.

When the war ended, there were 18 UDTs totaling 1800 men. By 1948 the number of teams were reduced to 2 comprised of 200 men. Still, frogmen had established themselves as an important part of the navy. Those who stayed in were dedicated frogmen. They continued to develop new techniques. UDT officers were sent to Europe to study the techniques of the Italian and British frogman units. And finally, several years after the war, American frogmen were given their first breathing apparatus: the Cousteau-Gagnan aqualung.

When the Korean War broke out in 1950, the navy found itself short of frogmen, and those the navy had were in the wrong places. The North Koreans had overrun most of South Korea. Now General MacArthur, commander of the United Nations forces, had the job of driving them back north. The only large seaport still controlled by the Americans and their allies was Pusan, but it was so over-crowded that MacArthur's landing force had to put ashore else-where. A beach seven miles from Pusan was chosen. First, a recon-naissance had to be made to find the safest places for the landing craft to enter. At this time, all of the UDT men were still in the United States. Rather than delay the landing until their arrival, military leaders gathered a makeshift unit of volunteers from the army, navy, and air force. They did the job, and soon afterward, UDT men arrived to take over the underwater work. Once again, frogmen had played an important role in military operations. During the war, in addition to reconnaissance and demolition, they were used for salvaging sunken ships and rescuing downed pilots.

When the Korean War ended, the navy resolved never to be caught unprepared again. This time, instead of cutting down on the UDT training program, they allocated more money for developing underwater techniques. Today frogmen have many uses. They par-ticipate in scientific research projects in the Arctic and Antarctic, recover manned space satellites, and undertake salvage operations, among other things. The training recruits receive in the two under-water demolition schools at Little Creek, Virginia and Coronado, California is the most rigorous of its kind in the world. All of them are volunteers, and more than 70 per cent are weeded out before they complete the first five weeks of basic training. Although the United States started late, today it has the best trained and equipped frogmen in the world. The British and Italian navies now send men to American UDT schools.

Within the naval amphibious forces at the present time are two types of navy frogmen, UDT personnel and SEAL team personnel. SEAL teams are navy units trained to conduct unconventional or paramilitary operations and to train personnel of allied nations in such operations. "SEAL" is derived from "sea, air, and land." Two SEAL teams have been formed as components of the UDT units at Little Creek, Virginia and San Diego, California. Under the "am-phibious force commanders" for training and administration, they are deployed to conduct naval operations.

Basically SEAL training is the same as the UDT training required for preparing beaches for amphibious landings. In addition to work-ing with surface ships, SEALs, like UDTs, are trained to operate from submarines and can be air-dropped into coastal areas. Their training differs from UDT training, in that they are expected to operate with little support and in restricted waters or on land. Because of this, their training is designed to give them the ability to survive in shallow waters or ashore, wherever they may encounter enemy forces. They are trained in hand-to-hand combat, the use of antipersonnel weapons, survival techniques, and special languages, training that usually isn't required for UDT frogmen.

The development of SEAL teams from underwater demolition teams was a natural extension of U.S. Navy amphibious capability as the need has increased in recent years to give training to the indigenous forces of friendly countries. For example, SEAL teams were used in the Vietnam War.

10

SUBMARINES

Man's drive to explore the sea by means of a closed container has led to another important invention—the submarine, a vehicle that can not only submerge and surface under its own power but which can also propel itself underwater. The principal reason for the development of such a device was for use in warfare.

The earliest mention of a self-propelled undewater vehicle occurs in the writings of the Swedish historian, Olaf Magnus, who wrote that during an expedition of the Swedish King Haakon to Greenland in the early sixteenth century, two sealskin submarines, each carrying three men, were captured. The natives had built them because they wanted to revenge themselves on European traders for their dishonesty. The Greenlanders bored holes in the bottoms of a few European ships before being captured. These submarines were propelled by oars jutting out through their sides, but because the sealskin hulls couldn't withstand the increased water pressure at greater depths, they weren't used at depths of more than a few feet.

The Englishman William Bourne, who designed the first diving chamber and wrote a book in 1578 entitled *New and Possible Inventions,* earned immortality for himself by dreaming up what is generally considered to be the first workable submarine. His submarine consisted of a wood and leather vessel that could be submerged and rowed, even though Bourne was somewhat skeptical of why anyone would ever want to row underwater when it was so much easier to do on the surface.

It is possible to make a shippe or boate that may goe under the water unto the bottome, and so to come up againe at your pleasure. Any magnitude of body that is in the water, if that quality in biggnesse, having alwaies but one weight, may be made bigger or lesser, then it shall swimme when you would, and sinke when you wish: and for to make anything doo so, then the jointes or places that doo make the thing bigger or lesser must bee of leather; and in the inside to have skrewes to winde the thing also outside againe: and for to have it sinke, they must winde the thing in to make it lesse, and then it sinketh unto the bottom: and to have it swimme, then to winde the sides out againe, to make the thing bigger, and it will swimme according unto the body of the thing in the water.

In 1620 a Dutchman living in England, Cornelius van Drebbel, built a submarine similar to Bourne's, which worked. The hull was covered with leather treated with tallow to make it watertight. For propulsion there were 12 oars, six on each side, emerging from the hull through watertight leather joints. When his submarine proved successful, Drebbel built an even larger one, adding more oars. Neither submarine ever went below 12 feet, but they were capable of making four miles an hour. For more than 10 years they went up and down the Thames almost daily, drawing observers from all over Europe. Several thousand people went for a ride, among them King James I who made an underwater trip of a few miles. Drebbel attempted to sell his invention to the British Admiralty for use in warfare, but it was rejected.

Then in 1653 a Frenchman named De Son decided to build a submarine at Rotterdam. In a statement he made in attempting to find backers, he claimed, modestly, that his invention could "go from Rotterdam to London and back again in one day, and in six weeks go to the East Indies, and to run as swift as a bird can fly. No fire, nor storm, nor bullets, can hinder her, unless it please God. Although the ships mean to be safe in their havens, it is in vain, for she shall come to them in any place and destroy them. . . ." The power of advertising being limited in those days, De Son didn't find backers, so he had to build his submarine largely at his own expense. It was 72 feet long and eight feet across the beam. The material was wood strengthened by iron girders, and large iron rams were constructed fore and aft for use in destroying enemy shipping. Propulsion was provided by a large paddle wheel in the center of the submarine, which extended into the water. A clockwork mechanism was supposed to turn it. In tests made while the sub was on the ways, the device provided propulsion for eight hours before it had to be rewound. As soon as the sub entered the water, however, it was found that, due to the water's greater density and weight compared with that of the air, the clockwork mechanism wasn't strong enough to move the paddle wheel. The submarine was a failure. In order to regain some of the money he had spent on the project, De Son turned the sub into a sideshow, charging admission to anyone curious enough to want to inspect it.

Inventors must have been discouraged by De Son's failure, for nothing was heard of submarines for over a century. In the June 1749 issue of *Gentleman's Magazine,* there appeared an article about a submarine, which included a drawing. The vessel in the drawing was virtually identical to Drebbel's machine, but Drebbel wasn't even mentioned. The author of the article, a Frenchman named Mariotte, took full credit for the idea. He claimed that he could regulate the depth at which the submarine could move but didn't say how. Without further details (it's assumed the submarine wasn't built) we can't consider Mariotte's submarine an advance over Drebbel's.

During the seventeenth and eighteenth centuries there was occasional mention in the press of submarines being designed and built, but few documents have survived, describing either them or their capabilities. In 1772 a Frenchman named Le Sieur Dionis is supposed to have built a submarine capable of carrying six people on a four-and-a-half-hour underwater journey. This wasn't likely, since, unless Dionis had a secret way to refresh the air, the submariners would have died from anoxia.

The distinction of building the first submarine to be used in warfare belongs to an American, David Bushnell. In 1776, as the Revolutionary War loomed, Bushnell was sent by the governor of Connecticut to General Washington to tell the general his plans for building a submarine that could seriously damage the British fleet. Washington listened with interest and immediately realized the shock value of such a weapon. He was keenly aware that the Americans had no navy to speak of. Writing to influential people and requesting help, Washington saw to it that Bushnell got the money to turn his idea into reality.

Bushnell's egg-shaped submarine, the *Turtle,* was constructed of wooden barrel staves and iron bands. For submersion, water was brought into the ballast tanks. The *Turtle* was propelled by means

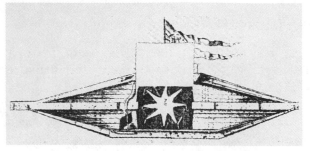

De Son's submarine. The eight-pointed star in the center is the paddle wheel that failed to work in water.

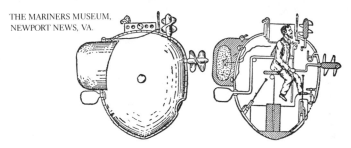

Bushnell's Turtle. Note the number of gears and levers within reach of the pilot's hands.

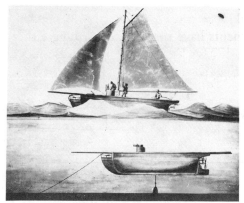

Robert Fulton's submarine submerged and under full sail. Plate VII from Fulton's "Drawings and Descriptions."

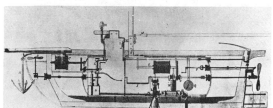

Cutaway view of Robert Fulton's submarine. From the first plate in "Drawings and Descriptions."

USN

C.S.S. H. L. Hunley (1863–64) *at Charleston, South Carolina, December 6, 1863.*

USN

The U.S. Navy's first submarine, U.S.S. Holland, *invented and designed by John P. Holland, at Elizabeth, N.J., 1898. The* Holland *carried a six-man crew.*

of a primitive screw propeller on the front; the speed was regulated by the man operating the propeller. Another screw propeller was mounted on top, for ascending. At the back was a rudder which was controlled by another operator. On top were the conning tower and two short tubes, one for intake and the other for exhaust. The tubes remained above the surface until the submarine submerged. Normally, it cruised underwater, with the upper part of the conning tower above the surface. The operator looked through the viewing port while he steered. When it was necessary to avoid being sighted by lookouts or when contact had to be made with the lower part of a ship's hull (to attach a 150-pound magazine of powder which worked on a delayed-action fuse), the submarine could operate completely submerged for short distances.

A few days before the signing of the Declaration of Independence, a large British fleet appeared along the eastern seaboard off the coast of New York. It was a formidable show of force. Washington, knowing he had no means of giving battle, decided to harass the British with the *Turtle* and ordered Bushnell to get it ready. Bushnell chose the flagship of the fleet, *H.M.S. Eagle*, a 64-gun warship guarding troop transports which were carrying 20,000 soldiers for an attack on New York City. But before the *Turtle* could be prepared for action, the British landed at Gravesend Beach on Long Island. Meanwhile, Washington's small army, dug in on Manhattan, was enduring fierce bombardment from British warships offshore.

Ready at last, the *Turtle* was towed from New Rochelle. At the helm was a sergeant from Connecticut, named Ezra Lee. A few miles from the British warships in the East River, the tow was released and the *Turtle* began to drift with the outgoing tide toward the British. Lee maneuvered close to the *Eagle* without being sighted and submerged. It was the only luck he had that night. Bushnell had given him a drill to use in penetrating the *Eagle's* hull, so that the explosive charge could be attached. But Lee soon discovered that the wood was sheathed with heavy copper (a recent innovation on British warships, the sheathing was to prevent the bottoms from being eaten by teredos, the destructive sea borers). During that long night Lee tried again and again to drill through the copper. It was nearly dawn, when he realized that he had to give up. As the *Turtle* pulled away from the *Eagle*, however, she lost her trim and floated to the surface before Lee could regain control. Afraid that he would be sighted, Lee again submerged and headed full speed for the Manhattan shore. But his compass was off. When he surfaced to get his bearings, he found himself under the British redoubts on Governor's Island. An alert British sentry spotted him and sounded the alarm. Within minutes the British had launched a boat in pursuit. Lee refused to surrender and instead, released the explosive charge with its delayed-action fuse. The British, who might have been puzzled by this strange craft, knew a bomb when they saw one and abandoned the chase, turning back to Governor's Island. Lee reached the friendly shore at the Battery a few minutes before the charge went off, making quite a bang but harming no one.

A few days later, Lee attacked the British fleet again, this time aiming for a frigate which, according to the reports of spies, wasn't protected by copper sheathing. But he was spotted and had to beat a hasty retreat. Undaunted, Lee made still another attempt. This time he was not only spotted but fired on. By this time, Bushnell realized his secret weapon was no longer a secret to the British, and he called off the attacks, rather than risk having his submarine captured and perhaps used against the Americans. Dismantled and carried inland, the *Turtle* was a closely guarded secret until after the war. Bushnell later designed and built other submarines but found little interest in them in the new United States, probably because the *Turtle* hadn't exactly distinguished herself in battle. He even tried to interest the French government, to no avail.

In spite of this, however, Bushnell's submarines attracted the attention of other inventors, among them Robert Fulton, the inventor of the steamboat. This Philadelphia farm boy was in Paris, designing a submarine of his own at the time Bushnell was trying without success to sell his invention there. Apparently Fulton was a better salesman. The French decided to invest in his submarine, which as yet existed only on paper. Fulton obtained a contract from the French government and built his submarine at Rouen, christening it *Nautilus*. Ellipsoid in shape, the submarine was 21 feet long, with a beam of seven feet. The hull was copper reinforced by iron frames, and propulsion was by Bushnell's hand-turned screw propeller. There was a rudder, as well as a mast and sail for cruising on the surface. Like the *Turtle*, the *Nautilus* was designed to attack enemy warships by carrying powder magazines to be attached to their hulls. In 1801 the *Nautilus*, carrying a crew of three, was successfully tested at Le Havre. But for some reason, it didn't impress the French sufficiently for them to back it further.

By the 1830s most European navies were building and testing hand-propelled submarines, making them larger and larger and of stronger materials, in order to reach greater depths. By 1870, the year *Twenty Thousand Leagues Under the Sea* appeared, no fewer than 30 submarines had been built. Clearly, interest in them was growing.

After studying the hydrodynamics of porpoises, an inventive German named Wilhelm Bauer designed a submarine which he called the *Sea Diver*. Bauer convinced the German government to finance its construction. Built in 1849 at Kiel, *Sea Diver* was man-powered, with the eight-man crew turning the propeller. She had four viewing ports and a hand pump to evacuate water from the ballast tanks. During a test dive in 1851 the submarine dived into a 60-foot hole, which wrecked the controls and caused leaks. Bauer and his crew let the rising water compress the air inside so that it was equal to the water pressure outside. Then Bauer opened the top hatch, and he and his crew "went up like champagne corks." It was history's first escape from a sunken submarine. Because of the incident, however, the government decided against funding any more submarine development work, and Bauer faded back into anonymity.

The submarine got its baptism by fire in the American Civil War. Great changes in warfare had taken place since the Napoleonic wars. As far as submarine warfare was concerned, though, little progress had been made. It took actual conflict to necessitate new inventions. The fact that the Confederate states were so inferior to the Union in naval power was a great stimulus to using submarines and mines. With the North maintaining a blockade of Southern ports, Confederate leaders came to view the submarine as the only way to break the Yankees' stranglehold.

Actually, the first move in submarine warfare seems to have been made by the Northerners. Records indicate that early in the war the federal government entered into negotiations with an unidentified Frenchman for the construction of a submarine to be used against the Confederate monitor, *Merrimack*. The government offered to pay £2,000 for the vessel's construction and £1,000 each for any successful attack on Confederate shipping. But before anything concrete materialized, the Frenchman absconded with the first installment, leaving the government with nothing more than his plans. The plans did not seem to have attained even the standards of Fulton's submarine 60 years before. Details of its size and other features are lacking, but it was built. Before seeing any action, it sank in a storm.

The Confederates built several small submarines, called "David boats," presumably in honor of David Bushnell. Built from cylindrical steam boilers taken from Mississippi river boats, they were cigar-shaped, varying in length from 20 to 40 feet. The larger models carried a crew of nine. Propelled by the same kind of hand-turned screws the *Turtle* had used almost a hundred years earlier, they could make about four knots (nautical miles per hour). On its bow it carried an explosive charge, 90 pounds of gunpowder with a delayed-action fuse. In the time-honored fashion, the charge was to be attached to the hull of an enemy vessel. One of the David boats, the *C.S.S. Hunley*, had the distinction of being the first submarine to sink an enemy ship in wartime. Up to then, the *Hunley* hadn't distinguished herself; twice she had sunk during tests, drowning her crew each time. Then the *Hunley* was given the task of sinking, or at least damaging, the Yankee ironclad, *Housatonic*, which was part of the blockade of Charleston, South Carolina. At sundown on the evening of February 17, 1864, the *Hunley* attacked the *Housatonic*. The charge went off prematurely, and both ships were rocked by the tremendous explosion. Once again, the crew of the *Hunley* were lost, and this time, the sub with them. Only five aboard the *Housatonic* were killed, but the ironclad was lost.

The major mistake made during this attack—as, indeed, with numerous others during the Civil War involving these so-called submarines—was that the *Hunley* wasn't used submerged, but only in the "awash" position, thereby losing the advantage of concealment. No doubt the reason for this was the problem of air supply for the crew. Air was needed not only for the crew, but the engines that were put in later David boats when it was found that the hand-turned propeller screw was unsatisfactory. Although only one submarine was sunk during the Civil War, the war resulted in important development of the submarine. People continued to debate the morality of submarine warfare at the same time that they were becoming more convinced of its utility. At the end of the war, there was a lot of haphazard experimenting going on, then methodical, continued progress which has continued to the present time.

During the period when the David boats were causing military leaders in Washington to lose sleep, a new submarine was built in France—the *Plongeur*. She was the largest submarine (140 feet long, with a displacement of 420 tons) up to that time and the first to be powered. The *Plongeur*, which was the invention of a naval captain named Bourgeois, ultimately revolutionized naval warfare. The hull was entirely of iron, and the propellers were turned by compressed air. The sub could maintain a speed of five knots for several hours at a time. However, the ballast mechanism was poorly designed. The first deep-sea trial almost turned into a disaster when the crew had extreme difficulty surfacing. Rather than risk this again, the French declared the *Plongeur* unfit and converted it into a large water-storage tank. French inventors quickly set about improving the design.

The English beat them to it. In 1878 the Reverend G. W. Garrett designed and built the first steam-powered submarine, the *Resurgam*, and in 1886 J. F. Waddington went a step further. He designed a submarine christened the *Porpoise*, which used electricity supplied by batteries and a seven-horsepower motor for power. Thirty-seven feet long and seven and a half feet across the beam, the sub proved to have good underwater stability. She could reach creditable depths and was capable of an underwater speed of three knots.

Also in 1886, the French were back in the race, this time with the *Gymnote*, a submarine designed by Dupuy de Lome. Naval historians generally consider the *Gymnote* the first truly modern, practical submarine and the forerunner of those in use today. Cigar-shaped, with a steel hull 60 feet long, and 6 feet across the beam, she was powered by electricity, like Waddington's *Porpoise*, but attained better speeds—seven knots on the surface and five underwater. Submerged, the *Gymnote* had a range of more than a hundred miles.

England quickly caught up with France, and before long the United States and Germany were in the race. Everyone tried to improve on the *Gymnote*, making new models larger, stronger, and with greater range. In Liverpool, the Reverend Garrett and a Swedish engineer, Torsten Vilhelm Nordenfelt, developed the first submarine with true military potential. Garrett wasn't thinking of war when he began designing the sub; he merely wanted to explore the ocean depths. Nordenfelt emigrated to England in 1862 and became the owner of a gun factory which did a thriving business with the Confederacy. The two men joined forces in 1865 and built the *Nordenfelt I* in 1881—the first submarine capable of attacking an enemy ship with a self-propelled torpedo. Trials held in the Baltic Sea, 15 miles north of Copenhagen, provided the occasion for a Danish royal house party. The Danish royal family, the Czarina of Russia, the Prince and Princess of Wales, and more than 50 ministers and military experts from many countries attended. The 65-foot submarine was powered by a steam engine designed by Garrett. *Nordenfelt I,* carrying a four-man crew, could submerge and travel for 15 miles or so at a speed of three knots. A glass-domed tower, called a cupola, through which the captain sighted the target, remained above the water when the sub fired her torpedo. The inventors claimed that the torpedo had a range of at least 500 yards.

Damage to a horizontal fin limited the first day's test to surface runs and inspection by distinguished visitors. The next day, despite poor weather, the submarine made several short dives, as well as a mock attack on a surface ship. The crew were injured by the rough seas, and the actual demonstration of a torpedo being fired at a target (which was what most of the observers had come to see) never took place. Although the event received a lot of publicity, no major power showed any interest or submitted any bids for buying production models from Garrett and Nordenfelt. Some years later, they sold the submarine to the Greek government, and more extensive trials were held in Piraeus Harbor in 1886. These trials were said to be more successful, yet the submarine was never actually used. In 1901 it was cut up for scrap.

A Turkish military attaché had been present at the demonstration in the Baltic. When it became apparent that there would be war between Greece and Turkey, he convinced the Sultan to commission the building of two submarines of a newer design. Garrett and Nordenfelt built them at Barrow-in-Furness, England and shipped them in sections to Constantinople where they were assembled. The submarines outperformed the earlier ones in every way. Each sub was 100 feet long, 12 feet in the beam and could attain 10 knots on the surface and three submerged for nine hours at a stretch. It carried two quick-firing surface guns, could fire three torpedoes, and was capable of submerging to the surprising depth of 100 feet. The submarine *Abdul Hamid* was launched at Constantinople on September 6, 1886, and sea trials were held in February. When it was determined that the sub was very hard to control underwater, Garrett rushed back to England to make modifications on the other submarine being built. In January 1888, both submarines were ready for testing. They successfully fired torpedoes with remarkable accuracy. The subs were commissioned in the Turkish navy but were never used in combat. Because of Turkish secrecy, we don't know what became of them, except that they were still on the Turkish Navy's list 20 years later.

In 1895 the U.S. Navy awarded a contract to John Holland to build an electrically-powered submarine similar to the *Gymnote*. Along with another American, Simon Lake, Holland had already built successful, hand-propelled submarines. Due to unrealistic specifications set by the navy, however, the submarine Holland built was a failure, and the navy didn't purchase it. But Holland wouldn't accept this. Privately, he built another submarine, this one to his own specifications. Its acceptance by the navy in 1900 marked the beginning of America's modern fleet of submarines. Holland also started the Electric Boat Company which, among other things, built the world's first nuclear-powered submarine, *Nautilus*, launched in 1955.

Beginning in 1900, the U.S. Congress authorized money for the development and construction of submarines. In 1913 a squadron of five submarines was permanently stationed at the Panama Canal to guard against attack by enemy submarine. By this time, submarines were already considered a formidable weapon, and World War I left no doubt. German U-boats almost strangled England, a country heavily dependent on imports. In 1916, the first year the U-boats were used on a large scale, the Germans sank over six million tons of British shipping.

The Allies should have learned something from their submarine warfare experience in World War I, but they didn't. At the outbreak of World War II, the Germans were far out in front. They had a much stronger sub fleet. Once again, the United States and Great Britain were on the defensive and forced to launch attacks at short notice. Once they got started, though, their record improved. While the total number of British and American submarines was less than half that of the Axis (506 and 1,253, respectively), they accounted for almost as many sinkings (2,370 enemy merchant ships sunk, compared to 2,752). In addition, Allied submarine losses were much less, 128 versus 835.

After World War II every major power, aware of the danger of lagging behind, recognized the need for submarine strength, and the submarine race was on. The most notable advances in the U.S. fleet were the conversion from electricity to nuclear energy and the deployment of the Polaris missile. The bulk of the American submarine fleet remains the submarine that evolved from the *Gymnote.*

The new Trident undersea nuclear weapons system—with its long-range missiles, larger submarines, and complete U.S. "home ports"—has increased combat readiness and cut operating costs. The weapons system consists of Trident missiles which are capable of reaching an enemy anywhere on the globe, which has made expensive, overseas bases unnecessary. The missiles of the Poseidon class had a range of only 2,500 nautical miles, compared to the Trident's 4,000. The Trident submarine is much faster, it can dive deeper and stay submerged longer. It carries a crew of 150 who live in virtual luxury compared to life aboard a World War II submarine.

11

SKIN DIVING

With a thrust of one's flippers and a hiss of compressed air, a whole new underwater world has opened up for diving enthusiasts. Inspired by Cousteau's beautiful films, as well as "Sea Hunt," "Primus," and "Flipper" on television, some 200,000 Americans a year are taking to skin diving. Today there are between two and three million amateur divers in the United States. To accommodate them, diving shops and schools have sprouted like sea anemones. The Professional Association of Diving Instructors (PADI) has grown from 500 members in 1966 to 7,000 today, and the skin-diving industry does an annual $50 million business in manufacturing alone. A recent profile of underwater enthusiasts found them to be young (average age, 24) and affluent (income ranging from $10,000 to $25,000 a year). For those starting out in SCUBA diving, an initial outlay of $750 to $1,000 is required. For those interested in underwater photography, another $500 must be spent. In 1971 only 5 percent of the nation's divers were female; at present this figure is close to 25 percent. Some parents start teaching their children the basic techniques of diving when they are as young as five, many people who are in their sixties and seventies enjoy the sport, too.

Many luxury resorts are paying more attention to the vacationing diver. On the Red Sea, the Club Méditerranée has a club exclusively for diving. Underwater visibility there is a fantastic 200 feet, and the marine life is abundant. Resorts in the Caribbean, such as Cozumel Island in Mexico and the Cayman Islands, cater to skin divers. In fact, nondiving guests are becoming an endangered species. Determined to dive year round in all kinds of weather, many aquatic addicts forego diving in tropical waters and instead dive near home, sometimes under bone-chilling conditions. For instance, thousands of divers in the Midwest think nothing of chopping through several feet of ice in the winter, to dive in lakes and stone quarries. And New Englanders dive for lobsters and fish even where there's six feet of snow on the ground ashore.

However, there are a number of experts around these days who think that the number of divers and their activities should be restricted. These experts claim they're causing irreparable damage to fish and underwater archeological sites. Cousteau, who spent the first 10 years of his diving career in the Mediterranean spearing fish for a living, is a prominent advocate of banning spearfishing everywhere. He and those like Dr. Hans Hass, a well-known marine biologist, have argued before such august institutions as the United Nations that skin divers are annihilating coastal fish populations, that somehow an international law should be passed prohibiting the manufacture and use of spear guns. The islands of Bermuda, Bon Aire, and Antigua in the Northern Hemisphere, and the Seychelles in the Indian Ocean have already banned spearfishing. In many areas off the coast of Florida, spearfishing is allowed only beyond three miles offshore. A Florida law states that there is a fine of $5,000, plus confiscation of all diving equipment, for merely picking up an old bottle or sherd of pottery at a shipwreck site without official permission from the State of Florida. Meanwhile, in the State of Washington, the Department of Fisheries has passed a law forbidding skin divers to wrestle with octopi (a local pastime among the divers), because it feels that the octopus population has been decimated since these wrestling matches began some 20 years ago. Skin divers often use chemicals to lure the octopi, some of which are 20 feet long, from their caves. After an exciting test of strength, the divers release them, but quite often the octopi die from fatigue and shock. To fight pollution and help safeguard the seas, Cousteau has formed a politically active group called the Cousteau Society, which already has 175,000 members.

Credit for starting the sport belongs to an American writer living on the southern coast of France in the 1920s. In love with the sea for much of his life, Guy Gilpatric was very curious about what lay beneath the surface of the Mediterranean. Using flying goggles waterproofed with putty, he made his first dive in 1929. What he saw made him feel as though he had entered an entirely new world. Gilpatric spread the word to his friends, who soon fell in love with diving, too. It wasn't long before descents for observation became descents for fishing. Divers found that with spears, they could return bearing trophies fit for any gourmet. Spearfishing caught on all along the coast of France and then spread to other countries on the Mediterranean. So great was the number of people who took it up that commercial fishermen began to complain that their fishing grounds were being depleted. They and the skin divers have been feuding ever since, with the result that the sport is now prohibited in some parts of the world.

Skin divers at first used goggles but soon changed to the more comfortable (and practical) face mask. Snorkels appeared a short while later. Now the diver could swim with his head underwater. Freed from having to raise his head for air, he could devote his attention to scanning the depths for fish. With the invention of swim-fins by a Frenchman, Louis de Corlieu, in 1935, the gear of skin divers was complete.

The sport was slow to catch on in the United States, though. In 1939 an American swimming champion and sportsman, Owen Churchill, visited Tahiti where he saw some boys wearing swim fins shaped like the tail of a fish and made of soft crepe rubber stiffened by metal bands. Albeit crude, they enabled the boys to swim almost as fast as Churchill. The fins captured Churchill's imagination, and on his return to the United States, he applied for a patent on swim fins, only to discover that Corlieu already had one. So he went to see Corlieu and arranged to license Corlieu's fins for manufacture in the U.S. Churchill's own swim fins, made of hard rubber, were introduced in 1940, but skin diving was so little known in America that his sales for that year were a mere 946 pairs. During the war period, when the usefulness of fins in frogman operations was recognized, he sold more than 25,000 pairs.

Churchill got some of his friends in southern California interested in skin diving. A few began to appear on the public beaches where their masks, snorkels, fins, long underwear (for protection against the cold), and eight-foot spears made them look as strange as Martians to the ordinary swimmers. After the war, when their ranks were swelled by former UDT men who couldn't stay away from the water, skin divers became a familiar, if not always welcome, part of the California beach scene.

It was not until 1951 that the sport achieved national popularity, and then things seemed to happen overnight. Early that year, Hollywood released the film *Frogman*, depicting the heroic exploits of UDT men during the war; and Cousteau's excellent film, *The Silent World*, followed soon thereafter. When the filmmakers saw the response that the films received, they realized they were onto a good thing, and many others appeared, such as *Creature of the Black Lagoon* and *Underwater. Underwater* was a $3 million extravaganza shot in Technicolor and financed by Howard Hughes. It was an exciting movie about searching for and finding sunken treasure in the Caribbean. No one left the world premiere with dry eyes. But it wasn't a tear-jerker; the premiere was held 18 feet underwater at Florida's Silver Springs. The viewers wore SCUBA tanks and sat on benches to watch the film.

As a result of these films, interest in skin diving spread rapidly. Dozens of companies sprang up to meet the demand for equipment. In 1952, when Cousteau put his "Aqua-Lungs" on the market in the United States, they were snapped up as fast as they could be produced. A Los Angeles firm that sold only a thousand spear guns in 1950 sold 280,000 in 1953 and had a six-months backlog of orders. And there was a magazine devoted to skin diving; copies vanished from newstands minutes after they appeared. In 1951, when *Frogman* was released, there were about 50,000 skin divers, most of whom lived in southern California. By late 1953 there were over a million in the United States. They formed clubs with such names as "the Neptunes," "Sea Urchins," "Mudsharks," "Diving Kids," "Sea Fools," and "Bottom Scratchers." Although the sport attracted men who enjoyed outdoor sports, it wasn't limited to them. Women and children also took to it. For some of them, it was the only way to see their husbands and fathers on weekends.

In September 1953 *Newsweek* ran an article on the fast-growing sport, which said, in part:

> Along many an American shore this summer, innumerable persons emerged from the water or vanished into it looking like webfooted Cyclopean apparitions. Round glass masks, covering eyes and nose, gave

them that one-eyed look. On their feet were big rubber flippers. Usually they carried spear guns, some patterned after the Medieval crossbow. Sometimes they were encased from head to foot in rubber suits and had oxygen bottles strapped to their backs. They called themselves skin divers, or spearfishermen, or goggle fishermen, or frogmen, depending on where you found them, and this summer they could be found in almost any American spot that had interesting water.

In areas where fish abounded, spearfishing dominated the activities of the skin divers. In August 1953 the first annual national underwater spearfishing championship was held around the kelp beds off California's Catalina Island. The championship was won by the "Muirmen" of Pasadena, whose three-man, four-hour total of 134 pounds of edible fish topped the "Miami Neptunes" who came in second with 78 pounds. The event was nationwide, with three teams coming from Puerto Rico. The next year the championship was held on a beautiful reef 12 miles off Key West. It was won by the "Miami Tritons." During the contest a diver shot a jewfish that weighed 210 pounds, a record, but it was the only fish his team got. Six months later this record was broken by a 403-pound catch and a month after that by one weighing 942 pounds, which was speared off the coast of Brazil. Records were being broken all over the globe, but some people—namely, rod and reel fishermen, weren't very enthusiastic about this. In fact, many were downright hostile toward skin divers.

Armed with spear guns powered by heavy rubber bands or surgical tubing, springs and carbon dioxide, skin divers prowled the waters, with deadly effect on fish. Sometimes, however, they injured each other. At first, commercial fishermen, as well as rod-and-reel sportsmen, viewed them merely as pests. But after a number of articles written by fishermen appeared in national magazines, stating that the nation's waters would soon be fished out by skin divers, there was open conflict in some fishing areas, especially along the coast of Florida. In Tampa Bay, commercial fishermen began snagging skin divers with grappling hooks and in their nets, and the divers retaliated by occasionally shooting a spear through the bottom of a commercial boat. In an incident off Jacksonville, a skin diver who was snorkeling under a public pier was killed by a large stone thrown by a fisherman. Another time, a skin diver was shot to death in the Florida Keys by an angry commercial lobsterman who accused the diver of robbing his lobster traps. The dispute became so serious in Florida that several city and county governments imposed a moratorium on spearfishing in their jurisdictions. It was some years before peace was restored.

In the early days of skin diving the public was concerned about the possibility of sharks attacking skin divers, particularly the spearfishermen who swam around with the bloody fish they had just speared. As time went on and many divers reported friendly encounters with sharks, it became apparent that the vaunted killer of the seas was really a bluff and a bully; often he was a coward, to boot. Skin divers soon learned that when they shouted or waved their arms underwater, sharks would usually take off. This included the infamous "killer sharks" such as the tiger, hammerhead, mako, and great white shark. Adding insult to injury, a California skin diving club had a membership requirement for which applicants had to free dive with only mask and flippers—no weapons—and wrestle a shark to the surface. (It must be said that the sharks were usually one of the gentler varieties, such as nurse, shovelnose, or sand shark.) Florida has the most heavily shark-infested waters in the Northern Hemisphere, rivaling the Great Barrier Reef of Australia. Yet, in the 25 years or so since the skin diving craze began, only two or three skin divers have been killed, and there have been only about a dozen attacks by sharks in which a diver was injured. Dr. Perry W. Gilbert, one of the world's best-known shark experts, says: "A very liberal estimate of shark attacks in the world would be 100 per year, and less than half are fatal. In fact, more people are killed by bee stings and lightning bolts in the United States each year than [are] killed by sharks throughout the world."

Understandably, many divers were, and are, still wary of sharks. It was finally decided that some kind of protection was needed for them. Research showed that shark repellants used in World War II weren't very effective; in fact, they sometimes *attracted* sharks. In 1955 a device called the "Electrical Shark Repeller" was introduced, which was similar to an electric cattle prod. But the Shark Repeller proved too effective: as frequently as not, it shocked the diver along with the shark he was trying to frighten away. Soon after this, the "bang-stick" was invented and is still the most widely preferred shark deterrent around. The original bang-stick was simply a long pole with a waterproof shotgun shell on one end, which was fired by putting it against a shark's head. This usually left a gaping hole in its head, killing it. The bang-stick had one drawback: it could fire only one shell at a time. When several marauding sharks were in an eating frenzy, the diver had some anxious moments, for it took two to three minutes to extract the spent cartridge and reload. A former bullfighter turned marine biologist solved the problem in 1972 by inventing a five-shot bang-stick. But the year before, a scientist at the Naval Undersea Research and Development Center in San Diego had devised the "Shark Dart" which is silent and, unlike the bang-stick, causes no bleeding. The latter is important because blood attracts other sharks, often triggering an orgy in which sharks attack everything in sight. The shark dart uses a blast of carbon dioxide to inflate the shark, which leaves him floating belly up. Such an injection can kill a shark, provided the dart penetrates the inner cavity of its body. Nowadays, most divers who spend any time in tropical waters carry a dart strapped to one of their legs. While a diver may never use it, it does give him a psychological boost.

Actually, more divers have lost their lives by drowning in the hundreds of sinkholes dotting the Florida peninsula than from shark attack. While most divers are content to spend their time prowling for fish and lobsters or capturing their beauty on film with underwater cameras, others have been lured to the mysterious, pitch-black sanctum of blind catfish and see-through shrimp, either in search of prehistoric fossils, bones, and Indian relics or for the thrill of the danger in exploring the underground labyrinths. The Jenny Springs sinkhole has claimed 11 men so far, three of whom drowned during one dive. They foolishly went down without a safety line and with only one flashlight among them. The flashlight burned out, which caused them to become lost and thus run out of air. In a sinkhole nearby, four university students drowned during a dive in 1967.

Since 1960 at least 180 skin divers have died in these sinkholes, and many others had such close calls that they gave up diving—at least in sinkholes. Over 90 percent of the deaths have occurred on the diver's first dive into a sinkhole or underwater cave. Because of this, the National Association for Cave Diving was founded by a group of Florida skin divers. The main purpose of the association is to teach basic underwater safety to inexperienced cave divers.

According to a study at the University of Rhode Island, in 1972, 139 Americans were killed while diving. In 1973 this number jumped to 151, seven of them women. California led with 34 deaths by drowning, followed by Florida with 31, of which 18 were the result of diving into sinkholes and caves. Twenty-four deaths were caused by diving with no breathing equipment; but most of them were probably caused by diving too deep and holding the breath too long (to do so causes temporary blackout, in which case one drowns). Only nine professional divers were killed in diving accidents in 1973, however.

In 1974 two teenagers died from nitrogen narcosis brought on by diving in 150 feet of water. When the Los Angeles County supervisors heard about this, as well as the fact that the boys were inadequately trained, they investigated. Then reporters from the *Los Angeles Times* learned that a national diving association had certified a dog and a seal as diving instructors through a mail-order program, and that, according to the captain of a charter diving vessel, many certificates issued to divers in southern California were worthless, that some of his passengers could barely swim and didn't have the foggiest notion of the proper diving techniques. Worse still, at Catalina Island, the most popular diving area in California, in a week's time there were three deaths—even with "certified" instructors present. During an 18-month period more than 40 divers had died, and in nearly every case, they were poorly equipped and apparently inadequately trained. Some had no training at all. The supervisors tried to get the county to ban further diving instruction and to stop the issuing of diving certificates until they could study the entire situation. But many leaders in the field of diving instruction, such as John Gaffney who runs the National Association of Skin Diving Schools (NASDS), fought them hard, and eventually the instructors won their battle. The county did, however, pass an ordinance setting better guidelines for SCUBA instruction. The matter soon blew over when some mediocre SCUBA instructors were forced out of business.

Some states now have laws which state that SCUBA divers cannot get their tanks refilled with compressed air from a dive shop or other commercial business unless they hold a proper SCUBA certification card from a training academy or diving school such as NASDS, YMCA, or PADI. Unfortunately, there is no law prohibiting a diver from buying an air compressor and recharging his own tanks. As John Gaffney says: "Only an idiot would try to fly an airplane

Neal Watson and John Gruener at a depth of 170 feet on their way down on their record dive.

LEFT: *Skin diver with playful dolphins.*
RIGHT: *The author's wife Jenifer with a large red snapper she speared.*

Diver Marcia Denius surfacing with two large prehistoric shark teeth.

without proper training and some supervision under the watchful eye of a qualified instructor, and the same applies to SCUBA diving. Only idiots or persons with suicidal tendencies are foolish enough to dive without adequate training." The typical SCUBA training program costs between $50 and $150 and consists of 10 to 20 hours of training in classrooms, swimming pools, and the sea or a lake, usually culminating in a supervised ocean dive to 100 feet. Before anyone attempts to go deeper than 100 feet, he or she should have at least 50 hours of diving experience in open waters, and then do so only with someone who's been diving for years.

By far the most dangerous skin-diving activity is the competition to set depth records. Spurred on by the accounts of Cousteau's group, as well as others who dive for scientific purposes and under optimum conditions, some skin divers have attempted to reach new depths for no other reason than the desire for publicity. In 1953 a 58-year-old Miami lawyer named Hope Root attempted to break Maurice Farques' record of 396 feet, set in 1947 on a dive that cost him his life. Diving with an insufficient supply of compressed air in his double-tank aqualung, Root reached 400 feet, then probably had an attack of nitrogen narcosis. He kept going down and was never seen again. After his death a number of scientific articles appeared, warning about the dangers of using compressed air below 200 feet. In spite of this, however, some divers have foolishly ignored the warnings. In 1955 John Clark-Samazan attempted to reach the 400-foot mark. At 300 feet he, too, blacked out from nitrogen narcosis but managed to reach the surface alive.

In October 1967 Kitty Geisler set a depth record for women by diving to 325 feet. She was accompanied on the dive by Neal Watson who had decided to try to break the men's record while he was at it. At the time, the men's depth record, using compressed air, was 390 feet, held by Hal Watts of Orlando, Florida. No man using SCUBA had dived beyond 390 feet and lived. Now Watson, who ran a skin-diving business in Freeport, the Bahamas, was determined, not only to go past this fearsome barrier, but to try to reach 450 feet! All of his diving acquaintances tried to talk him out of it, but he was determined. He even talked another diver, John Gruener, into diving with him. Unlike most of the divers who attempted depth records, Watson and Gruener trained for nearly a year, making deep dives every day and scientifically planning for every contingency they could think of. On dives beyond 250 feet, they usually experienced nitrogen narcosis. About the effects of nitrogen narcosis, Watson said:

> Although we were flirting with death on each of our deep practice dives, we never failed to become fascinated by the sensations we experienced each time we probed the depths. Sometimes I felt like I was a king who owned the whole world; other times I would be traveling to the moon on a flying carpet or have hundreds of women chasing after me. Usually right after the narcosis would hit us we would stop the dive, however; the moment we turned upwards and started for the surface, we would either lose consciousness or become so intoxicated that we didn't know up from down. We finally devised special safety features which would make our dives safer. We would lower a line to the bottom with a hundred-pound weight. A bolt was inserted in the line at the maximum depth we planned to reach on each dive. Each of us had a short length of line attached to one wrist, which was also attached to a shackle on the diving line. This served three purposes: we would both descend and ascend together, we would be prevented from swimming away from the diving line when intoxicated by narcosis, and the line would effectively stop our descent. We also held onto a short slip line to which a sixty-pound weight was attached, so we would be pulled rapidly towards the bottom. If either of us blacked out, he would automatically drop his end of this line, the weight would be released and both divers would float slowly to the surface because of emergency buoyancy floats which we attached to the back of our diving tanks. If all went well on the dive and we were both still conscious when we reached maximum depth, we would release our weight, attach our clips and swim or climb back up the line. The perfection of this system took many months of dives, some of which almost ended our ambitions to go for the record.

Before attempting the record dive, Watson and Gruener broke Watt's depth record unofficially on many of their practice dives. But they took the precaution of having a standby safety diver equipped with a helium-oxygen mixture to assist in an emergency. On October 14, 1968 they went for the record. After ascending, neither of them could remember anything after they had reached the 400-foot level. There were some anxious moments while the officials authenticated the dive. When the weighted line was pulled up, they saw that Watson and Gruener had attached their clips at 437 feet one inch —a record on compressed air which still stands. Both men now say that breaking the record was exciting, but that they wouldn't do it again. Since Watson's and Gruener's feat, there have been eight attempts to break their record—all of them fatal.

Besides spearfishing, there are other activities pursued by divers. Where there are only a few fish, skin divers are usually content to dive merely for the sake of diving. Many have taken up underwater photography, while others explore submerged caves. After some divers were killed when they got lost in the caves and ran out of air, however, cave-diving was left to the very proficient.

Still other divers pursue treasure, guided by maps and books which purport to give the locations of vast riches throughout the Western Hemisphere. These guides aren't cheap, either. Many would-be treasure hunters have learned to their sorrow that the job should be left to professionals who rely on authoritative information. The only treasure that amateurs consistently find is in the form of gold nuggets, relics of prospecting in rivers of the western states, particularly California, which fell into holes when the gold was washed down-river.

Many skin divers use their favorite sport to make some extra money. They dive for such items as anchors, fishing equipment, outboard engines, even golf balls lost in water holes on golf courses. One Floridian is averaging $50,000 a year recovering some 500,000 golfing mistakes. Probing the muddy bottoms of the water holes, he fills a large burlap bag with balls on each dive. Recently, near Traversham, England, SCUBA divers solved a two-year-old crime by discovering seven pieces of sixteenth-century silverware valued at over $3,000. The pieces had been stolen from a nearby church, and their find led to the capture of the thief. And on the seafloor off Key West, skin divers found 25 sacks of marijuana worth $100,000 that must have been jettisoned to avoid capture.

In 1965 the Outboard Marine Corporation introduced a diving rig with a surface-air supply system (SAS), which was far superior to any others. It is for use in shallow waters by amateurs who aren't interested in getting into SCUBA but who still want some means of breathing underwater. The same type of unit is distributed by Johnson Motors as the "AIR BUOY" and by Evinrude Motors as the "AQUANAUT." The SAS unit weighs only 40 pounds and consists of an engine-pump assembly held up on the surface by an inflatable tube. A two-horsepower, two-cycle engine delivers two and a half cubic feet of clean air per minute to each of two divers who operate from the same unit up to a depth of 25 feet (the length of the hoses). This is twice the amount of air needed at this depth. Also, by attaching additional lengths of hose, many divers have descended to 50 feet. Depending on the carburetor adjustment, the unit will run for about 60 minutes before requiring more gasoline and oil. The cost of operation is about 15 cents an hour, which is cheap compared to refilling a SCUBA tank (for shallow water, about $2 an hour). The SAS unit has several unique features. The air is very clean and "tasty," unlike the dryer air one breathes from a SCUBA tank. The hoses don't kink or sink to the bottom and become tangled. And the full face mask fits every face, whether child or adult. The units are excellent for teaching people the rudiments of diving, especially children, because the length of hose keeps them from going too deep where they might get in trouble; also, since they're attached to the surface unit, it's easy to keep track of them. Boat owners also find them useful for making minor repairs underwater, cutting the lines out of a propeller, and cleaning the bottoms of their boats. The units have also been a hit at Caribbean resorts where tourists with no background in SCUBA diving can safely poke around the beautiful reefs. During my three-year excavation of the sunken city of Port Royal, Jamaica, my divers and I used them for a total of 13,000 hours underwater without a single problem.

12

UNDERWATER ARCHEOLOGY

All over the world there are objects lying on the bottom of the sea which belonged to civilizations long since vanished. Like relics painstakingly unearthed by land archeologists, those found underwater are clues to the past. In 1928 Salomon Reinarch, a leading Hellenist, wrote: "The richest museum of antiquities in the world is still inaccessible to us. It lies at the bottom of the eastern Mediterranean. We are able to explore the land and air without much difficulty, but we are very far from rivalling the fish in their element, which in the words of St. Augustus, 'have their being in the secret ways of the Abyss'. Those ways remain closed to us." Reinarch made this statement before the great steps were made which have enabled man to begin to discover the great treasures of the undersea museum.

The first people to show interest in underwater archeology were a group of English antiquarians who in 1775 sponsored an expedition to recover historical artifacts from the Tiber River near Rome. Using a diving bell, Greek divers worked for three years with little success. They had no way of removing the river mud that had been accumulating for centuries, and which covered some wrecks they wanted to search for artifacts. Interest in underwater archeology declined after that, and it wasn't until early in this century, when Greek and Turkish fishermen and sponge divers brought up some objects, that the interest of archeologists was again aroused. When the archeologists saw how beautiful and valuable many of these items were, they hired divers to search for more of them. Most of the magnificent bronzes now on display in the National Museum in Athens were recovered by these divers.

Fishermen have, no doubt, been finding unique hand-crafted antiquities for centuries, many of which must have been melted down for scrap. Early in the eighteenth century, fishermen near Livorno, Italy recovered bronze statues of Homer and Sophocles, and shortly after that, in the Gulf of Corinth, they snagged a bronze statue of the "Zeus of Livadhostro." In 1832, off the coast of Tuscany, near the site of the ancient Etruscan city of Populonium, fishermen hauled up a bronze statue of Apollo. This exquisite work is today one of the Louvre's major exhibits. Until the mid-1960s, when a dredging operation at Piraeus uncovered many spectacular bronze statues, the Apollo bronze was the only bronze original from the Greek Archaic Period (before 480 B.C.).

In 1900 sponge divers reported finding a large cache of statuary at a depth of 30 fathoms off the island of Antikythera. A salvage operation to the site was sponsored by the Greek Ministry of Education. Due to the depth, however, the divers had only about five minutes' bottom time. This was emphasized when one of the divers exceeded the time and died from the bends. It wasn't a proper archeological excavation, either. The uneducated divers had no idea what was worth saving and what wasn't. Several large marble statues which they mistook for boulders were mistakenly picked up by cranes and thrown into deeper water. The site was not worked systematically, and when the exhausted divers stopped, there were still many valuable objects remaining on the bottom. The wreck proved to be a Roman argosy laden with works of art that had been looted from Greek temples. Amphoras and other ceramic objects found at the site date the wreck at about 75 B.C. In addition to a number of priceless bronze statues, including the well-known "Antikythera Youth," 36 marble statues and a bronze bed decorated with animal heads were raised. Smaller objects such as a gold earring in the form of Eros playing a lyre, some exquisite glass vessels, and the gears from a unique astronomical computer were also brought up. This discovery provided magnificent examples of original works by master sculptors of the Argive and Athenian schools. French divers who briefly revisited the site in 1953 claim that the wreck itself is buried deep in the sand and that many more works of art await recovery.

Another Roman argosy dating from the first century B.C. was found by Greek sponge divers in 1907, three miles off Mahdia on the coast of Tunisia. The bulk of the ship's cargo was stone bases, capitals, and 60 columns, each weighing about 200 tons. Between 1908 and 1913 divers worked on the site, which was in 150 feet of water. Under the direction of the Tunisian Department of Antiquities, they recovered most of the cargo, which included well-preserved bronze busts, dancing dwarfs, a heron, and a large statue of Agon. In 1948 a team of French divers led by Cousteau used SCUBA equipment for the first time on an underwater archeological site. With water jets, they removed the overburden of mud covering the wreck and raised more of the ship's cargo to the surface. The site was again worked in 1954 and 1955 by amateur divers from the Tunisian Club of Underwater Studies, but there is still much to be done there.

In 1928 sponge divers made still another accidental discovery off Cape Artemision in Greece. There the divers found a Roman shipwreck dating from about the time of Christ. Employing helmet divers, The Greek navy conducted a salvage operation which ended abruptly when a diver died from the bends. The famous statue of Zeus (or Poseidon, the god of the sea), a bronze jockey, and parts of a galloping horse of the Hellenistic period were among the objects found. Although the wreck lies in sand only 130 feet down, nothing more in the way of proper excavation has been done.

The first *disciplined* archeological work on an underwater site was done in Lake Nemi near Rome in 1928. Legends had grown up about two Roman ships that sank there during the first century A.D. Both were enormous and sumptuous. Over 230 feet long, with decks paved in mosaics and colored marble, they had heated baths, marble columns, and other luxurious features. The ships were probably pleasure craft for the Roman nobility. There had been feeble attempts to salvage them in 1446 and again in 1535, but the equipment used was too primitive. Twice during the nineteenth century, divers recovered artifacts from the site. The last operation until this century was halted by the Italian government in 1895 when it was discovered that divers were removing large quantities of wood planking from the wrecks.

Then in 1928 Mussolini decided that the government of Italy should salvage the ships, and over a period of four years, the lake was drained. When the ships were exposed, and before they were disassembled and taken to a warehouse in Rome, archeologists had a rare opportunity to study the two well-preserved hulls. Fortunately, they made good, detailed drawings, for the ships were burned by the Germans in 1944. Today we have only scale models to study.

The first major excavation of an underwater site in the Western Hemisphere was led by Edward Thompson, the American consul at Merida, Yucatán, early in this century. Passionately interested in archeology, Thompson purchased some land containing the ruins of Chichén Itzá, the most important city of the Maya. Under the auspices of the American Antiquarian Society and Harvard University's Peabody Museum, he began a systematic exploration of the site. Near one of the temples, he found a large *cenote*, a hole formed in the limestone when the roof of a subterranean cave collapsed, which is filled with rain water that has collected. Thompson's *cenote* was 190 feet in diameter and the water in it was 65 feet deep. The walls rose 60 feet above the water's surface. Local Indians told him that the *cenote* was called the "Well of Sacrifice." According to legend, it contained an immense treasure. Thompson's research confirmed the Indians' story. For centuries before the Spaniards invaded Central America, the Maya had worshipped Yum Chac, a rain god believed to inhabit the *cenote*. Dependent on the maize crop for food, the Maya made offerings to Yum Chac, especially during times of drought. These offerings included gold and jade and occasionally, young maidens.

Returning from the United States in 1909, where he had raised money from the Peabody Museum and learned to dive in a helmet suit, Thompson began work. A derrick was built for lowering and raising him, as well as holding a suction dredge to be used to remove the mud from the bottom of the *cenote*. Thompson's early dives revealed that much of the mud on the bottom was 10 feet deep and that the water was inky, it was so dark. For months Thompson relied solely on the suction dredge, which was manipulated from the surface. Tons and tons of mud were pumped up, but the ooze didn't yield even one artifact. Just as he was about ready to admit defeat, the dredge brought up the first one: a round ball of resinous incense. Thompson then switched to a steel-jawed grab bucket. After that, hardly a day passed without a relic being unearthed. Thompson dived into the gloomy depths, working by touch, and there found more incense balls, ceramic incense burners, vases, bowls, plates, axes, lance points, arrowheads, copper chisels, discs of beaten copper, even human bones. Among the artifacts were about $800,000 worth of gold bells, figurines, and discs, as well as pendants, beads, and earrings of jade. But Thompson wasn't particularly interested in making money; he was an amateur archeologist. So the priceless collection, like many others in the early days of archeology, was shipped to the Peabody Museum where it was put on display until 1960 when it was given to the Mexican government.

During the Revolutionary War several British warships were sunk in the York River, just off Yorktown, Virginia. In 1934 oystermen discovered the hulks of some of them, and the Colonial National Historical Park Service and the Mariners Museum of Newport News joined in salvaging them. U.S. Navy divers, using water jets to blow the mud off two of them, reported that they were too poorly preserved to be raised. Therefore, a grab bucket was used. Operated from a barge, it recovered a fairly representative collection of late-eighteenth-century armament and equipment—cannon, anchors, weapons, ship's fittings, tools, bottles, crockery, and pewterware. During the summer of 1976 Dr. George Bass, who had recently formed the American Institute of Nautical Archaeology, spent six weeks surveying the shipwrecks. After locating 12 possible sites with a magnetometer, he dug test holes at some of the sites and found many interesting and important artifacts. Lack of money has prevented him from continuing his work, but he hopes to raise the necessary capital soon and continue the excavation.

The grab bucket Thompson used was still in use in 1950. The techniques and standards of underwater archeology had progressed little in 40 years when Nino Lamboglia, director of the Institute of Ligurian Studies, initiated a project which pointed up the destructive capabilities of the grab bucket. In 1925 fishermen had snagged amphoras in 140 feet of water off Albenga, Italy from what proved to be a first-century B.C. Roman shipwreck. Lamboglia could not get government funds for a salvage project, nor could he arouse any interest among amateur divers. Thus, when he was offered the assistance of a commercial salvage firm, he accepted. Under his direction, helmet divers removed a few of the amphoras by hand. Then a large grab bucket was used, which was directed by an observer in a diving chamber who was in telephone communication with the surface. Giant steel claws smashed into the wreck, wrenching up amphoras, wood, and other objects. The excavation went on for 10 days, during which 1,200 amphoras were raised. All but 110 of them were broken. Such destruction of an archeological site was to stimulate the development of scientific techniques for future underwater archeological projects. Lamboglia was the first to admit that he had made a serious mistake in not making drawings of the site or laying better plans for the excavation.

The next major excavation of an ancient wreck took place in 1952. Divers working on the construction of the great Marseilles sewer outlet off the small island of Grand Conglouré come upon the remains of a second-century B.C. Roman ship in 150 feet of water. Excavation of this wreck became the proving ground for many of the tools and techniques used today.

Captain Cousteau joined forces with Fernand Benoit, a land archeologist for the project. To clear away the layer of mud and sand that covered most of the site, Cousteau used a kind of underwater vacuum cleaner called an "airlift." Airlifts were used by treasure hunters in the Florida Keys during the 1930s, but never before had they been used on an archeological dig. Benoit and the other nondiving archeologists watched the operation over closed-circuit television. They were frustrated, however, because they couldn't communicate with the divers and thus exercise better control over their activities. And the divers, in turn, were limited by the depth of the water to three dives a day, or a total of 45 minutes of bottom time. Fortunately, while several thousand dives were made during the project, only one diver lost his life in an accident.

Although many unusual and priceless artifacts were recovered, the excavation wasn't considered a total success by many archeologists, because of the failure to gather pertinent data while the artifacts were still on the seafloor. There was no plan of the site, and neither Cousteau nor Benoit provided information on how they had arrived at their conclusions about the wreck site. It's a well-established rule of archeology that archeologists record and publish their findings, for what may seem unimportant at the time can prove significant in the future. However, like Henrich Schliemann, the excavator of Troy and Mycenae, these men were pioneering a new field, and their mistakes should be judged accordingly. Due to other commitments following the project at Marseilles, Cousteau and his team could not devote themselves to underwater archeology for many years. In 1968

Cousteau made a brief expedition to Silver Shoals, off the coast of Haiti, in search of a Spanish galleon that had sunk in 1641 with a large treasure in gold, silver, and gems. Due to faulty research, however, they failed, having to settle, instead, for photographing the wreck of a Dutch merchantman which sank during the eighteenth century and which had been thoroughly worked by treasure hunters over the years. Then in the summer of 1975 the Greek government hired Cousteau to make an archeological survey in the Aegean Sea. When he and his team didn't find any interesting shipwrecks, they decided to explore a known wreck off the island of Antikythera— the Roman wreck found and salvaged by helmet divers at the turn of the century. They recovered three gold bars, a large number of gold coins, and an engraved jewel-studded necklace.

The first complete and successful excavation of an ancient shipwreck to be directed by a professional archeologist working underwater took place at Cape Gelidonya on the coast of Turkey. The project was the idea of an enthusiastic American named Peter Throckmorton—erstwhile student of archeology, diver, and a man in the throes of a wanderlust that had brought him to Turkey after roaming all over the world. In 1959, after sharing countless bottles of *raki* with some garrulous Turkish sponge divers, Throckmorton learned the location of areas containing "old pots in the sea"—amphoras. Nearly every sponge diver in the Mediterranean knows of a wreck site or two, and during the course of the next year, Throckmorton checked out about 35 of the areas that the divers had mentioned; most of them where ancient ships had sunk over a period of some 2,000 years.

One of these wrecks, lying in 90 feet of water off Cape Gelidonya, turned out to be the second oldest shipwreck ever found, a Bronze Age wreck dating from 1300 B.C. Throckmorton realized its importance as soon as he had made an exploratory dive. He convinced some sponge divers who had been planning to dynamite the site and raise the cargo of copper and bronze to sell for scrap, to leave it alone for the moment. Then he reported his find to the University of Pennsylvania's Department of Archaeology. Dr. Rodney Young, director of the university's Institute of Classical Archaeology, offered to find the money and personnel to mount a major expedition to the site.

In the summer of 1960 a team of 20 specialists in various aspects of underwater archeology arrived to join Throckmorton at Cape Gelidonya. The team included Frederic Dumas, who was experienced in underwater projects off the French Riviera, and Dr. George Bass.

Because of the wreck's depth, a diver was limited to 68 minutes working time per day. But since each man's task was carefully planned, the team was able to make the most of the brief bottom time. When the area had been cleared of seaweed, the team made drawings and plans of the wreck. Then they made a photographic mosaic of the site. On the bottom the site looked like a conventional land site, with meter poles staked about and numbered plastic tags marking the objects that were visible. Each item was drawn, photographed, triangulated, and plotted before being raised to the surface. A thick deposit of lime as hard as concrete covered most of the cargo, with only an occasional piece protruding through the hard seafloor. To try to extract individual pieces underwater was too time-consuming and risky, due to the fragility of many of the artifacts. Working with hammers and chisels, the divers broke off large clumps of conglomerate, some weighing 400 pounds, and sent them to the surface in lifting baskets where they were broken apart. Some of the larger masses had to be separated with automotive jacks. Until each day's recovery could be processed in this fashion, it was stored in a freshwater pond on the beach.

When the clumps were broken apart, a vast array of artifacts were found: bronze chisels, axes, picks, hoes, adzes, plowshares, knives, spades, a spit, and many copper ingots. An airlift being used in the few sand pockets at the site yielded other artifacts, including four Egyptian scarabs, oil lamps, polished stone mace-heads, apothecary weights, pieces of crystal, mirrors, awls, a cooking spit, whetstones, olive pits, and the bones of animals and fish.

Although little of the hull had been preserved, the distribution of the cargo suggests that the ship was approximately 12 meters long. Brushwood dunnage like that described in *The Odyssey* was still lying over fragments of planks. The cargo and personal possessions that were found indicated that the ship was a Syrian merchant vessel that had picked up its cargo of metal on Cyprus. The wreck shed new light on seafaring during the Bronze Age and furnished a wealth of information on early metallurgy and trade.

The expedition opened the way for future underwater archeological projects in the Mediterranean. George Bass, convinced of the importance of this fledgling science, decided to make it his life's work. Before he had left for Turkey, he had consulted several land archeologists and found that most of them felt that underwater archeology was impossible and could never become an exact science. Bass heard such reasons as "nothing could be preserved underwater" and "it's impossible to make proper plans underwater." Some said it was too dangerous and far too expensive for the amount of knowledge that could be gained.

During excavation of the wreck, Bass and the others had been able to prove these dire predictions unfounded. They discovered that a surprisingly large part of the cargo was in an excellent state of preservation despite having lain underwater for some 3,300 years. They had made accurate plans and drawings underwater, more professional, in fact, than many produced for land excavations. Also, although most of the expedition's members had little previous diving experience, there were no diving accidents. Finally, to the surprise of the skeptics, the whole project, including the air fare for the people involved, had cost less than $25,000. The only disappointment Bass and Throckmorton felt was that so little wood from the original ship had been preserved; this made it difficult to determine any details of the ship's construction.

The two men have worked together during nearly every summer since then continuing to make discoveries on the coast of Turkey. In September 1975 Throckmorton found a shipwreck even older than the one from the Bronze Age. At a depth of 75 feet, near the entrance to a secluded harbor near Hydra, an island south of Athens, he found the remains of a wreck dating from 2700 to 2200 B.C. Pieces of storage jars and other ceramic ware found there seem to indicate that the ship was a trading vessel. Unfortunately, the Greek government refused to grant Throckmorton a permit to excavate, and he has since returned to the United States where he plans to continue his work on American shipwrecks.

Within a year of the completion of the excavation by Bass and Throckmorton, the most important and challenging underwater archeological project ever undertaken in Europe was completed after five years of work. On April 24, 1961 the 64-gun Swedish warship *Vasa* was raised. The ship had sunk in Stockholm Harbor in 1628 in full view of the king and thousands of spectators shortly after being launched, carrying with her 50 passengers and crew. The deepest part of the wreck lay in a hundred feet of water—too deep for early divers working out of bells to accomplish much. Salvaging a wooden ship is usually impossible after 50 years; by that time, the wood has been eaten away by shipworms. Stockholm Harbor is one of the few places in the world where there are no shipworms, however, so the *Vasa* had remained intact all that time.

In 1956 a Swedish petroleum engineer named Anders Franzen became interested in the *Vasa*. After several years of research, he had narrowed down her location, and, using a core sampler (a device geologists obtain samples of sediment from the seafloor with) operated from a small boat, had finally found the wreck. Divers from the Swedish Navy were sent down to identify and investigate its condition. As Franzen had expected, they found it intact. With the encouragement of the Swedish government and money from private sources, Franzen began the tedious job of raising the ship. He used helmet divers, but the job was far from simple. The *Vasa*, large for her time, displaced 1,400 tons (four times the size of the *Mayflower*) and was buried deep in the mud of the harbor. Divers first had to remove all of the loose objects aboard. Then they blasted tunnels under the wreck with a water jet—a perilous undertaking, because the ship could easily have slipped deeper into the mud and crushed them—so steel lifting cables could be inserted under the hull. Constructing the tunnels alone took three years. While this was going on, other divers removed the masts, spars, and rigging. Finally, the cables were strung through the tunnels, and pontoons were used to lift the *Vasa*.

She was then towed into dry dock and placed on a specially built concrete barge equipped with a sprinkler system to keep her wet until she could be properly preserved. The moment the *Vasa* broke the surface, Franzen was very busy. With scores of archeologists and historians, he entered every accessible part of the vessel. They found themselves in a fantastic time capsule. Everything lay as it had fallen nearly three and a half centuries earlier: sea chests, leather boots, weapons, carpenter's tools, beer steins, cooking implements, money, powder kegs. More than 1,000 artifacts were found. Franzen even discovered the remains of some of the victims of the disaster, lying among the cannon carriages. Twelve partially clad skeletons were dug out of the mud inside the ship. The sheath knife and a leather

The Vasa *standing on a specially constructed pontoon in a dry dock. Most of the excavation work was done with the ship here.*

MARX

Diver using a metal detector underwater.

MARX

*Scenes from the author's large-scale excavation of Port Royal, Jamaica.
Divers are shown using the "AQUANAUT" surface-air supply system.*

MARX

*Diver holding gold coin near three large silver bars from a Spanish
shipwreck.*

money pouch containing 20 coins were still attached to the belt of a seaman. The *Vasa* was a veritable underwater Pompeii. Here, mud and cold water had taken the place of volcanic lava and ashes.

Since 1973 underwater archeologists have been exploring the remains of the U.S. brig *Defense*, a 16-gun vessel lost in Penobscot Bay, Maine during a battle with the British in 1779. She was one of 38 ships which entered the bay to attack the British-held Fort George. So far, divers have been occupied with mapping the site, and few artifacts have been recovered. Meanwhile, in Rhode Island a group of 18 cadets from West Point are assisting archeologists in exploring two other Revolutionary War shipwrecks, the British frigates *Cerberus* and *Orpheus*. The frigates were scuttled off Aquidneck Island in August 1778 to prevent them from falling into the hands of the French. The archeologists and cadets have recovered many valuable artifacts, and more excavation is planned for the future.

Underwater archeology ranges from excavating shipwrecks to excavating sunken cities, ports, and ceremonial wells. It includes sites once occupied by prehistoric man. Because there are less than 50 qualified underwater archeologists in the world, the future of this branch of science depends in large part on encouraging the motivated amateur diver to acquire the basic tools necessary for archeological research and excavation. Money for training such professionals is severely limited; thus, in the meantime, it is the amateur diver —the person for whom the underwater world is an avocation—who must fill the void.

13

SUNKEN TREASURE

Since the introduction of SCUBA diving in the United States, rarely has a year passed that some major underwater treasure find hasn't been announced in the press. The lucky salvors were assumed to have become instant millionaires, which is more myth than truth. Once a diver has caught a bad case of "treasure fever," it's difficult to cure. Otherwise sensible businessmen have given up lucrative jobs, sold their property, and headed for the sea armed with phony, or at least questionable, treasure maps and in a matter of months lost their life's savings. Others have learned of a shipwreck location from some fisherman in a waterfront bar. On many occasions, after receiving a handsome fee, the fisherman couldn't relocate the shipwreck, claiming that it had either been covered by shifting sands or "some pirate divers had already got to it and recovered everything off of it." In many cases, divers found a shipwreck, but after months of work and at great expense, ended up with nothing but a few worthless relics to show for their effort. Either the ship had already been salvaged or it wasn't carrying much when it sank.

A few amateurs have struck it rich, but for each one of these, a thousand have failed miserably. Even those who do find a large treasure have it taken by one government or another. Take Paul and Max Zinica who gave up their small ice cream factory in Gary, Indiana in 1967 to go treasure-hunting off the coast of Texas, looking for treasure galleons lost there in 1553. After searching for several months, they found one of the wrecks and brought up over $600,000 in treasure and artifacts. Then the State of Texas moved in and confiscated the booty, leaving the brothers only a single piece of eight. Today the two unlucky treasure divers are back in the ice cream business fighting a lawsuit with Texas, trying to get some of their treasure back.

With stories of successful recoveries of sunken treasure to spur them on, many people have spent years and untold sums of money pursuing treasure that never existed. A wild goose chase lasting for centuries involved the *Florencia,* a ship of the Spanish Armada, which supposedly went down in 1588. According to legend, the *Florencia,* carrying $5 million in treasure, was forced by bad weather and lack of fresh water to put into Tobermory Bay in Scotland. There the Spaniards took a hostage, a highland chieftain named Donald McLean who, in retaliation, set fire to the Spaniards' ship. The fire and resulting explosion killed everyone aboard, including McLean. A few months later, treasure-hunters descended on the wreck with diving bells. They found nothing, but for most of the next four centuries, hardly a decade passed when a search for the *Florencia* wasn't undertaken. The money spent in these endeavors must by now exceed the value of the treasure allegedly lying in the wreck. Ironically, it was confirmed a few years ago that, not only was the wreck not carrying treasure, it wasn't even the *Florencia.* The *Florencia* was one of the few Armada ships that returned to Spain. Experts are fairly sure that the wreck at the bottom of Tobermory Bay is that of another ship in the Armada.

Another wild-goose chase focused on Vigo Bay, Spain. In 1702, during the War of the Spanish Succession, a homeward bound Spanish treasure fleet of 20 galleons and its escort, 23 French warships, fought a combined fleet of English and Dutch warships. Legend has it that most of the Spanish ships, laden with treasure worth more than $50 million in modern currency, sank in Vigo Bay, which was thus paved with riches for those brave enough to seize the treasure. Many *were* brave enough to try. Since 1702 at least 75 treasure-salvaging expeditions to the site have been mounted. A few silver coins, valuable only to collectors, have been recovered.

Like the stories about Tobermory, those about Vigo are largely myth; for the record tells a different story. When the Spanish treasure fleet sighted the English and Dutch, it hurried into Vigo Bay where the Spaniards immediately began unloading the treasure (an amount only a fifth of what the legend claims). It was then taken overland to Madrid. But before the transfer had been completed, the

Anglo-Dutch fleet entered the harbor and the battle began. Eleven Spanish ships were captured, and the rest were burned to prevent their falling into enemy hands. Documents in the Spanish archives reveal beyond a doubt that no appreciable amount of treasure was on the ships set afire by the Spaniards. What wasn't taken to Madrid was captured by the English and Dutch.

The only treasure at the bottom of Vigo Bay was on an English ship, the *Monmouth*. Leaving ahead of the main fleet with the English share of the Spanish treasure, it struck a rock and sank in 150 feet of water. Because of the depth, no attempt to reach it was made until this century. In 1955 some American divers formed a company, raised $100,000, and went after it. Four years of continuous effort resulted in their finding what they believed to be the *Monmouth*, but there was no treasure on her. Did they find the wrong wreck? Or had the English managed to unload the treasure from the *Monmouth* before she sank? We don't know. Whatever happened, it's just one more tale of a fruitless treasure hunt to add to the many about Vigo Bay.

Through the centuries there have been innumerable quests for sunken treasure that existed only in someone's imagination. Even today, when people are supposedly more realistic, there's a thriving business in maps, articles, and books on treasures and their locations. Each year, thousands of gullible souls use them to hunt for sunken treasure in hopes of becoming rich overnight. But more money has been spent in the pursuit than has ever been recovered from the sea. Still the search goes on, and men continue to dream. Ask anyone bitten by the treasure bug why he hunts, and the chances are he'll say: "If William Phips could do it, so can I."

Ten years ago, people searching for sunken treasure were considered harmless crackpots who foolishly chased after a mirage. If they wanted to throw away their money, that was their business. Today, however, treasure-hunting is a big business. The latest scientific know-how and technology are used. Many companies spend as much as $10,000 a day. Mel Fisher spent three years and over $700,000 locating a galleon in the Florida Keys, which was lost in 1622 with about $25 million in treasure aboard her. Although most major finds are by professionals like Fisher, most of these professionals started out as amateurs with less knowledge of diving and treasure-hunting than the average diver has today.

When today's treasure-hunters were still wearing diapers, Art McKee was already a successful treasure-hunter. Although the years are beginning to catch up with him, he's still actively engaged in the profession. As the oldest and most experienced treasure-hunter in the business, McKee has a steady stream of visitors, most of whom are would-be treasure-hunters who come to him with the thinly veiled objective of picking his brain to learn where the rich treasure wrecks are. When asked such questions, McKee usually just smiles and says: "If I knew where, I'd be out on the wreck myself, not wasting my time talking to you!"

McKee's first contact with the sea was as a lifeguard at several New Jersey resorts during his summer school vacations. As a young boy he read a lot of books on sunken treasure and had got the treasure bug early. Upon graduating from high school McKee decided to become a deep-sea diver. He got a job as a diving tender for an old helmet diver who was repairing a bridge that had been destroyed by a hurricane. One day when the old diver was too drunk to dive, McKee talked a friend into tending the lines and made his first dive in a helmet diving suit. With no training, he was, naturally, afraid at first. But within a few minutes he had fallen in love with diving and hasn't stopped yet. His employer then put McKee to work as a full-time diver, and he quickly became a good one. After the bridge job was finished, he was hired to search for a large anchor that had been lost by a tanker in the murky waters of Delaware Bay. By accident he discovered the remains of an early eighteenth-century English merchant ship. Although all he found were pieces of broken rum bottles and ceramic sherds, he decided then and there to go after a real treasure wreck someday.

Longing for clearer and warmer waters in which to continue his diving career, McKee in 1934 moved to Florida where he took a job as recreation director for the city of Homestead. On weekends he prowled around the many reefs of the Florida Keys, searching for brass and other metals from modern shipwrecks. It was a pastime that was quite profitable. There were very few professional divers in those days. McKee was sought out for so many jobs that he finally had to quit his job as recreation director and go to work full time as a diver. People would come to him, asking if he would use his skill to find sunken treasure. Most of them were crackpots. There was an exception once, however:

I was approached by a man who really had a good lead from an old chart pin-pointing a treasure wreck. He offered me diver's wages for a ten-day job, and I accepted. When we got to the wreck site, which was in shallow water, I went down and discovered several cannons on the coral reef. Never having seen a wreck in clear water before, I just figured that the cannon had been jettisoned by a ship which had run aground on the reef, to lighten itself and get off again, but was I ever fooled. I spent the whole day probing around the reef without finding anything. Just when I was about to quit for the day, I noticed that down in deeper water there were more cannon laying about.

The next morning I dropped over the edge of the reef into eighty feet of water and landed on a sandy bottom. There were a lot of cannon strewn about. Noticing what appeared to be a large coral growth, I walked over and hit it with my pick. After knocking off a piece of coral, I saw a hollow place where there had been barrel staves, long since rotted away. I hit the coral a couple of whacks, and this exposed some hard pitch or tar with several shiny gold coins sticking to it. I broke off a larger piece of the stuff and sent it topside. They told me over the phone that they had found eighteen doubloons in it. Man, was I excited. I had only *dreamed* about gold. I'd never actually seen it on the ocean floor. After several hours of working, I broke the mass apart, and we found over sixteen hundred gold doubloons altogether. Unfortunately for me, since I was only on a diver's salary, my employer just gave me a few of the coins as souvenirs. I've been back to that same wreck at least fifty times since then but haven't found any more treasure.

It was a common practice in the sixteenth, seventeenth and eighteenth centuries for treasure to be concealed in this manner, usually to smuggle it into Spain to avoid paying the exorbitant import taxes and as a precaution in case the ship were boarded by pirates.

While working on a pipeline job in 1937, McKee was approached by an old fisherman who told him about a "pile of stones with corroded pipes on its top." On his first dive McKee saw that it was an old wreck. He brought up several coral-encrusted iron cannonballs and four heavily sulphated silver pieces of eight. Not sure what the coins were, he took them to a marine biologist at the University of Miami who told him they were worthless pieces of lead. But McKee wasn't satisfied with this answer and took them to a jeweler who said the coins were silver.

During his next dive on the wreck, McKee discovered a Spanish gold *escudo* stamped 1721, indicating that the wreck had occurred sometime after that. Exhausting every source in the United States and unable to learn the identity and history of this ship, he wrote to the director of the Archives of the Indies in Spain. Months went by. Finally McKee received a large package from Spain, containing hundreds of pages of old documents concerning the wreck he was interested in. In addition, there was a photograph of an old chart, showing the locations of 20 shipwrecks, one of them McKee's.

In July 1733 a hurricane struck a homeward-bound Spanish treasure fleet in the Florida Keys. All 20 ships in the fleet were wrecked on the reefs and shoals. Intensive salvage operations followed almost immediately after the disaster, and most of the treasure the ships were carrying was recovered. McKee's wreck was the most important ship in the fleet. Her name was *El Rubi,* but since she was the flagship under the command of Admiral Don Rodrigo de Tarres, she was also called the *Capitana.* Her original cargo consisted of treasure worth five million pesos. Unfortunately, according to the ship's manifest, it had all been salvaged.

For months McKee worked alone on the wreck. Anchoring his boat over the site, he would start his air compressor, don his shallow-water helmet, and go down a line to the wreck. There he would spend as many as 10 hours a day, moving the massive ballast stones and fanning away the sand by hand. There were interesting finds almost every day. During the summer of 1938 McKee found some good partners, which enabled him to step up the tempo of his operation. After the 250 tons-plus of ballast stones had been moved to one side, he realized that much of the cargo must be hidden deep in the sand. By hand, they could fan down only a few feet, so McKee decided to find a new tool with which to remove the sand. When he heard that navy divers were using a device called an airlift to deepen muddy harbors, he built one of his own. It was the first time an airlift had been used on old shipwrecks. Today, it's still one of the most important tools for underwater excavation.

During the next 10 years McKee and his partners searched the Florida Keys from one end to the other. They located and explored 75 shipwrecks, of which about 30 were old cannon wrecks. During World War II, when there was a great demand for scrap iron, commercial salvage divers raised most of these cannon, as well as anchors and anything else of value as scrap. McKee, realizing the historical and archeological importance of the items, tried in vain to stop them. By 1949 McKee had a warehouse full of artifacts and treasure from dozens of shipwrecks. On Plantation Key, only a few miles from the sunken *Capitana,* he built and opened to the public the first museum in the world devoted entirely to sunken treasure and artifacts.

The year before, Charles Brookfield, an old Florida "conch" who had begun to treasure-hunt shortly after McKee, told him that he had heard rumors that Bahamian fishermen had been finding large numbers of Spanish silver coins on the beaches of Gorda Cay (between Abaco Island and Nassau). Brookfield and McKee got up an expedition but couldn't find the shipwreck from which they were sure the coins were coming. They were able to buy some of the coins from fishermen, thus verifying the story.

They went back in 1949. While McKee walked the seafloor looking for the wreck, Brookfield tended his lines and searched with a glass-bottom bucket. Brookfield was luckier. One day he found two small ballast piles and signaled McKee to come over. As McKee was walking over, Brookfield, who was looking through the bucket, spotted a long dark bar. He hooked it with a long pole which the natives use for picking up conch on the bottom. McKee picked up the bar. It appeared to be iron, so he sent it topside with a line and came up himself. A few blows with a hammer revealed that it was a silver bar and that there were numerous markings on it. Brookfield asked his partner: "Are there any more down there?" and McKee replied: "Yes, one." "Well, get the hell down and send it up!" Brookfield shouted.

Before sending it up, though, McKee shot some movie footage of it, the first of an authentic treasure find. Then Brookfield went down and photographed McKee prying the bar loose from the ballast rocks, to which it was cemented by coral. Both ballast piles were resting on a limestone bottom. There was no sand to conceal any more treasure. They decided that if any more remained, it would be hidden in the ballast rock. Later that day, McKee found a third silver bar, this one weighing 75 pounds.

In the days that followed, the two men alternated shifts of moving ballast rock. Although they didn't find any more silver bars, they did find many artifacts, as well as about 50 silver coins.

Of all the underwater explorers I've known over the years, few have even come close to leading as exciting and interesting a life as Teddy

Tucker, who claims that he would prefer to have lived several centuries ago and been a pirate. Even today, he would be the perfect one to play a pirate in a movie. In addition to being one of the best-known residents of Bermuda, Teddy is one of the most knowledgeable and successful underwater treasure-hunters.

In 1938, at the age of 12, he convinced a helmet diver working in Hamilton Harbor to teach him to dive. Tucker became so fascinated with the underwater environment that he fashioned a diving helmet from a small boiler tank by connecting a garden hose to a hand-operated air pump on the surface. With some of his school chums he explored miles and miles of the reefs surrounding Bermuda. They earned spending money by selling coral, sea fans, and shells to tourists. Tucker loved the sea, and hated school so much that his parents had to physically take him to school each day.

At 15 he stowed away on a merchant ship to England. There he lied about his age and joined the Royal Navy, where he spent World War II. Tucker claims to have spent most of his navy years in brigs and jails. Once, while stationed at Plymouth, England, he was ordered to clean up the base's mascot—a jackass. He was given methylated spirits to clean the animal's hooves with. Instead, he mixed the spirits with beer and drank the mixture, ending up in the brig for several weeks. After the war Tucker supported himself by working as a commercial diver salvaging modern shipwrecks and other things. This gave him a chance to see the underwater world in such exotic places as the Malacca Straits, the Gulf of Siam, the Bay of Bengal, and other places in the Indian Ocean.

Returning to Bermuda in 1948, Tucker decided to forsake his reckless past and become a respectable citizen of the community. He even gave up drinking, which he blames for many of his past problems. With another diver, Bob Canton, who became his brother-in-law, Tucker started a commercial salvage firm in 1949. They recovered brass, lead, and other metals from modern shipwrecks. When the salvage business was slow, Canton and Tucker worked as commercial fishermen. He still makes a good living at this work during the winter months.

One day in 1950, while searching for one of his fish traps with a glass-bottom bucket, Tucker spotted two iron cannon on the bottom. A few days later he returned to the spot, which was about 10 miles outside Hamilton Harbor, and raised the cannon, as well as a large copper kettle full of lead musket balls. At first he and Canton were going to sell the cannon for scrap, but some members of the Bermuda Monuments Trust Commission heard of his discovery and offered him a lot more than he would have gotten by selling them for scrap. They went back to the wreck site and recovered four more cannon, an anchor, and a pewter plate. While the wreck was interesting, Tucker and Canton decided to stick to their normal salvage work.

One day late in the summer of 1955, after a storm had passed, the two men stopped at the wreck site, and Tucker jumped in with a face mask. The visibility underwater was excellent. He noticed that the storm had removed a great deal of sand. Reaching the bottom, he saw a piece of metal which he pulled out. It was a bronze apothecary's mortar, beautifully decorated, bearing the date 1561. Excited by his find, Tucker returned to the boat, started the air compressor, and jumped back in, wearing his shallow water diving mask. With a small board he began fanning the sand away in the area where he had found the mortar and in five minutes had a handful of blackened silver coins. He dug a trench about 18 inches deep. Soon a bright object fell out. It was a gold cube weighing about two ounces. Tucker was so excited that he bumped his head on the bottom of the boat while surfacing. Right then and there, he and his brother-in-law decided to forget about their salvage company and become full-time treasure-hunters.

They had about a week before bad weather would set in and force them to quit for the season. It was that week that they struck it rich, finding more gold cubes and larger gold bars, in addition to gold buttons studded with pearls and other gold jewelry. Then on the last day, Tucker discovered the most valuable single item of treasure ever recovered from an old shipwreck. It was a magnificent emerald-studded gold cross which was subsequently valued at $200,000. As might be expected, the find hooked them for good. Since then, they have become two of the most successful treasure hunters in the business and are still active in the field, despite the fact that they are in their fifties and sixties, considered advanced ages for a diver.

Unlike most of the other professional treasure hunters in the business, Kip Wagner got started rather late. But this had little effect on his success in the business. Wagner was the only treasure hunter who became a millionaire by recovering sunken treasure. He retired in the early seventies and died not long after that.

Wagner was born and spent most of his life in Miamisburg, Ohio. He claimed that he had absolutely no interest in treasure or shipwrecks until he moved to Florida after World War II. Shortly after starting a construction business, he began to hear stories about people finding gold and silver coins on the beaches, usually after storms. At first Wagner chalked them up as idle tales. One day a construction worker of his showed up dead drunk. Rather than fire the man, Wagner took him to the beach to let the fresh salt air sober him up. As they were walking along the beach, the man bent down and picked up what appeared to be a piece of rusty metal. It was a silver coin. During the next half hour, he found six more. He told Wagner he had found hundreds over the years.

Through research, Wagner learned that a fleet of 12 Spanish treasure galleons had been lost in the area in 1715 and that there was still millions of dollars worth of treasure lying in their remains. He got some local divers together and formed the Real Eight Company. Although they were amateurs, never having seen a shipwreck, Wagner and his divers learned the ropes by trial and error. He located the sites of most of the shipwrecks from the disaster in 1715 by finding coins on the nearby beaches. After exploratory dives in the spring of 1960, they got started. Wagner selected their first target. The task that immediately concerned them was to move the tons of ballast stones. They would have to do this by hand before they could reach the treasure. At first everyone worked enthusiastically, but after spending several weekends at the back-breaking work, they were skeptical. As they were approaching the point of calling the whole thing off, a diver found a large wedge of silver—and that was the beginning of one of the major treasure finds of this century. During the next 10 years, Wagner's Real Eight Company recovered over $10 million in treasure and artifacts.

It may sound preposterous that a man who has recovered millions in treasure from the sea has had the electricity turned off in his house because he didn't have the money to pay the electric bill, or that his wife has bought groceries for the family, using pieces of eight instead of dollars. But that's what happened to Mel Fisher. After a lifetime of successful treasure hunting, he's still trying to get rich. When asked what he did with all the millions he and his partners found, he answers quietly: "Man, most of it went to the State of Florida, my backers, and to Uncle Sam for income taxes." Actually, a lot of it was invested in expeditions to find shipwrecks. Fisher's favorite expression, which he repeats a dozen times a day, is: "There's a pot full of gold below; let's go get it."

He began diving at the age of 10, when he made his own diving helmet out of a five-gallon paint can. To sink it, he melted down his toy soldiers, but the helmet still had no window or air supply. His earliest dives were in a clear-water gravel pit. By looking down, he could stay down for about five minutes at a time. On a few occasions he stepped off a ledge and tipped the helmet, thus losing his air, and had to make a quick ascent to the surface.

After a near fatal accident in which he lost consciousness and almost drowned from rebreathing the stale air too long, Fisher drastically revised his helmet. He put a window in and fitted the valve stem from a bicycle tire on top. The helmet was connected to a bicycle hand pump via 15 feet of hose. A friend worked the pump while sitting on an inflated inner tube on the surface. On his first dive with the new helmet, in a lake near Gary, Indiana, Fisher had too many weights on and got mired waist deep in the silt on the bottom. To make matters worse, his buddy was pumping so hard that the glass window blew out, nearly drowning him before he could reach the surface. Again, he modified the helmet, putting metal bars over the window, and it worked well for many years.

Fisher got his first salvage job when he was 12. He and a friend raised a speedboat that had sunk in a lake. They bought it, minus the engine, for a dollar, added a sail and oars, and used it as a diving boat. When they were 14, they caught treasure fever from reading some books on the subject and decided to run off to Florida via the Mississippi River. They covered about a hundred miles when someone stole their boat and everything in it as they slept on a river bank. So they gave up and came home.

Fisher encountered skin diving during World War II when he was on furlough on the French Riviera. But he had no equipment and had to be content watching others spear fish and lobsters. After the war he spent a few years in the construction business in Chicago and Denver. His aim was to become a diver, however, and in 1948 he moved to Tampa and started his own construction business. The sport of skin diving was so new then that it was months before he saw anyone engaging in it. He was fishing from a bridge one day, catching a fish now and then, when along came a man with snorkeling equipment and a spear gun. Right where Fisher was catching one- and two-pound fish, the spear fisherman bagged a 60-pound grouper and minutes later a 35-pound snook. Fisher was so frustrated that he broke his fishing rod in two and threw it in the water. The next day he was back with his own equipment and a homemade spear gun. Until the diver who had been there the day before taught him the rudiments, however, it appeared that skin diving was harder than it looked. Within a few days he was free-diving to 50 feet and spearing fish with ease.

Fisher later moved to Los Angeles where he discovered what appeared to be an old cannon. As it turned out, this discovery caused him much embarrassment. News of the find created a lot of interest. It was the first cannon to be found in California waters, and the news media played it up big. NBC television covered it nationwide. It was placed on exhibition at a pier in Redondo Beach where thousands paid a dollar apiece to see it—until Fisher removed the coral encrustation and discovered that it was a sewage pipe!

When he failed to find treasure on the shipwrecks, Fisher tried diving for gold in California's rivers and streams. It began as a hobby, but when he made his first "strike" (actually, a few small nuggets), thousands of divers got gold fever, and Fisher turned to developing and marketing a jet venturi underwater vacuum dredge, of which he sold hundreds. To teach potential buyers how to use it, he began taking small groups to the rivers on weekends. The groups got larger and larger. One weekend he had some 500 families out in the mountains with him. Fisher says he never actually found much gold. He was too busy selling his dredges and other equipment. Some divers did find gold, though. One of them made $75,000 in a few weeks.

Sometime later, Fisher himself got treasure fever, making a number of unsuccessful expeditions to the Caribbean and coast of Panama. In December 1962 he was approached by one of Kip Wagner's divers and talked into joining the Real Eight Company team. He got together some close friends and formed the Treasure Salvors Company, and they headed for Florida. While the Real Eight team was picking up treasure hand over fist, Wagner had Fisher's team hunting for new wrecks. After a year of finding nothing, Fisher and his disgruntled associates were ready to go back to California. They had found a shipwreck and had been digging holes at it for weeks, without success. Then it happened: one of Fisher's divers discovered two seven-and-a-half pound gold discs.

The next morning Fisher said he had a feeling that if they moved about 500 feet farther seaward, they would make a big find. Everyone else thought it was crazy to move away from the area where they had found the gold discs, but the salvage boat was moved anyway. The blaster, an excavation tool used to remove the sand covering a wreck site, cut a hole 15 feet in diameter between two large coral heads, and divers went down to check it. They were almost blinded by the glow. The bottom of the hole was a veritable carpet of gold. The three divers stuffed coins into their gloves. When they dumped them in glistening cascades on the deck of the boat, the total came to 1,073. Mel Fisher's hunch had resulted in the most valuable treasure recovery of the century.

They returned to the area the next day, along with most of the Real Eight divers, and recovered another 900 coins. During the next five days another 600 were recovered, bringing the total to 2,573. But for the next few days, no more were found. Then the dry spell ended, and they began finding coins again.

During those days, Fisher says, he dreamed that the ocean floor was paved with gold and feared the world gold market would be ruined if word got out about their discovery. They adopted elaborate security measures, putting the treasure in various banks. But the news did get out. Then they had great difficulty keeping pirate divers away, especially at night when they weren't at the site.

By the end of the summer they had recovered over 3,500 gold and 6,000 silver coins, as well as numerous gold chains, medallions, rings, and other pieces of jewelry. The take was valued at over $1,000,000. To raise working capital for the next season, they sold 107 of the gold coins at auction for $50,000. Due to bad weather, that diving season was the worst they ever experienced. According to Wagner, without Fisher's dedication the season would probably have been a complete failure.

Treasure Salvors has continued to work with the Real Eight group each summer on the ships that sank in 1715. Together, they have made many other important and valuable recoveries. From 1971 to 1976, Fisher and his team searched the Florida Keys for the *Nuestra Señora del Atocha,* a treasure galleon they have now found and from which they are bringing up a portion of the millions on the wreck.

14

OCEANOGRAPHY

At the same time that diving leads to the past through underwater archeology and treasure hunting, it also leads to the future. The world's population is increasing dangerously, and the animal, vegetable, and mineral resources of the earth are rapidly being exhausted. In countries such as India and China, food supplies barely support the population now, with the situation becoming more alarming every year. The ocean and the land produce about the same amount of organic matter each year, but the share of sea products in man's food supply averages only 1 percent, or 70 million tons a year. In protein content this is equivalent to about 280 million head of cattle. Although the oceans are inhabited by more than 150,000 different animal and plant species, man thus far has tapped only 1,500. Scientists estimate that by the year 2000 at least 300 million tons of food must be recovered from the sea if man is to survive. Thus the logical solution is further oceanographic research to improve our harvesting of the sea. Much of this can be accomplished only by diving.

The oceans cover 140 million square miles of the earth's surface and occupy 324 million cubic miles, or 15 times as much as the land occupies. Some 10 million square miles of the seafloor lies on the continental shelves at an average depth of 600 feet. This depth is within the range of conventional SCUBA divers and those using mixed gases. The remainder of the oceans, which average 12,000 feet, lies within the realm of the exploratory capabilities of deep-diving submersibles, which have already shown that they can reach any depth and enable man to make important oceanographic discoveries.

SCUBA divers have investigated large portions of the continental shelf of the United States, as well as elsewhere; but thus far, they have been restricted to those areas where the water is no deeper than 250 feet, because mixed-gas diving has only recently been perfected, and its cost of operation is still prohibitive for most scientific organizations. Despite depth limitations in the past, SCUBA-diving scientists have accomplished much. Their most notable success so far is the discovery of underwater deposits of oil and natural gas; but other kinds of mineral deposits have been found as well, and plans are being made to begin mining them. Several American firms, for instance, are involved in dredging the unconsolidated offshore deposits of diamonds, gold, iron, and tin, in addition to lime sands, sand, and gravel. Millions have been spent, both on locating deposits of manganese nodules that also contain other scarce minerals and on the development of equipment for mining them.

The first known instance of undersea mining occurred 2,000 years ago when the Greeks dug lead and zinc ores out of the Mediterranean. For the last three centuries the British have been mining coal in subterranean tunnels. Today some 4,000 men work in these tunnels, providing 10 percent of Great Britain's coal requirements. Almost 30 percent of Japan's coal comes from under the sea. At the present rate of consumption, man will have exhausted the land deposits of many essential minerals within 100 years, resulting in a deficiency that can be made up only by undersea mining.

Many scientists feel that offshore underwater cities are one answer to the continually increasing pressure on land space by the world's expanding population. A British oceanographer recently suggested building an undersea city with homes for 30,000 people, which will offer all the amenities of life on land. Although this seems more like science fiction than a practical solution, some authorities view the idea as the only reasonable way to solve the problem. The government of Israel, for example, already has scientists at work designing such cities, and by the year 2000 they may be a reality.

Man's curiosity about the sea is as old as recorded history. For thousands of years men have been trying to probe and understand the oceans. Pytheas, an ancient Greek geographer, discovered that the tides are influenced by the moon's orbit. And that great philosopher of ancient Greece, Aristotle, catalogued and described 116 species of fish and 74 invertebrates in the Aegean. Posidonius, another curious citizen of the ancient world, braved the hardships of a voyage to Spain for one reason: to verify or disprove the belief that when the sun set in the western Mediterranean, it hissed and steamed like a red-hot poker as it plunged into the sea. During the Middle Ages, an Arabian naturalist named Masudi attempted to understand the mysteries of evaporation, the formation of rain, and the salinity of seawater. Centuries later, Columbus, Magellan, Cook, and others explored the sea and solved many of its mysteries. It wasn't until the early nineteenth century, however, that oceanography was recognized as a science in its own right. This was made possible through the untiring efforts of a U.S. naval officer named Matthew Fontaine Maury.

The first purely oceanographic data-gathering expedition took place in 1715 off the northern coast of Africa under the direction of a French diplomat named Benôit de Maillet. He employed divers to obtain information about tides, ocean currents, marine life, and geological formations. Not long after this, English natural philosopher Robert Boyle wrote several treatises based on his study of the ocean's depths, temperatures, and light penetration. During the eighteenth century, more than a hundred scientists in Europe studied the oceans and wrote about their findings.

Benjamin Franklin is considered to be the first American oceanographer. Throughout his active life he was inquisitive about the oceans and made several notable contributions to the study of them. His most important oceanographic work was in establishing the existence of the Gulf Stream, "that river of the Atlantic linking the Old and New World together." Franklin was also interested in diving. At an early age, he invented swim and hand fins. He reminisced about this in a letter in 1772:

> I made two oval palettes each about ten inches long and six broad, with a hole in the thumb, in order to retain it fast to the palm of my hand. They much resembled painters' palettes. In swimming I pushed the edge of these forward, and I struck the water with their flat surfaces as I drew them back. I remember I swam faster by means of these palettes, but they fatigued my wrists. I also fitted on my feet a kind of sandals: but I was not satisfied with them, because I observed the stroke is partly given by the inside of the feet and ankles, and not entirely by the soles of the feet.

Matthew Maury was the first to view the sea as an entity that gathers diverse phenomena into a single force. He methodically gathered information on the movement of the wind, the changes in climate, on tides and currents from mariners and scientists, and used it to devise a series of history-making "Wind and Current Charts and Sailing Directions for Seamen." Although most of Maury's research dealt with the surface of the sea, it provided the impetus for later study of what lies beneath the sea.

A British Admiralty expedition to the Arctic in 1819, led by Sir John Ross, spurred public interest in the sea as more than just a means of transportation and a source of food. For that day and age, Ross's research was remarkable. From depths of 6,000 feet, he brought up tube worms, a large sea star, and other marine organisms. Soon other British scientists were probing the depths of the sea with long sounding lines. When telegraph cables were laid across the floor of the Atlantic, their research took on a practical significance.

When it was learned that there is such a thing as deep-sea fauna, many scientists focused their attention on this subject. In 1862 Alphonse Milne-Edwards, an Englishman, published the first scientific study of deep-sea fauna. The science of marine biology had begun. In 1869 British scientists successfully dragged a trawl in the Atlantic at a depth of 14,595 feet over a seven-mile stretch of the seafloor. They hauled up a considerable number of unknown marine animals. One of the scientists present was Sir Charles Wyville Thompson. His appetite whetted, Sir Charles wanted to trawl every seabed in the world. He talked the Admiralty into backing his expedition, and they provided a steam-driven corvette, *H.M.S. Challenger.*

The *Challenger* sailed from Plymouth in December 1872 and stayed at sea for 42 months, collecting valuable oceanographic data. The scientists aboard her learned that the deepest point in the Atlantic is 22,500 feet and that the average depth of the Atlantic is 15,500 feet. In the Kuriles Trench off Japan, this record was broken when they dredged up marine life from a depth of 26,865 feet. The science of oceanography was in its infancy when the *Challenger* left England; but when she returned in 1876, oceanography was an established science. The 50-volume study that was subsequently published, the "Challenger Reports," remain the oceanographer's bible.

While bringing up marine fauna, fossil, and geological specimens was important, it was also necessary for scientists to observe them in their natural environment. This led to the development of undersea photography and deep-diving submarines. In 1855 Wilhelm Bauer made the first attempt to obtain underwater photographs. And in 1872, Major Daudenart, a Belgian, wrote a treatise in which he stated that photography could be successfully accomplished by mounting a special optical periscope on a submarine. Nothing, however, came of this. Then in 1875 an American named Eadweard Muybridge reportedly took the first good underwater photographs "beneath the waters of San Francisco Bay." Unfortunately, little is known, for the photographs haven't survived. Five years later, another American, George C. Moore, conducted experiments in underwater still photography offshore of Sturbridge, Massachusetts. The distinction of taking the first black-and-white photographs underwater goes to the biologist, Louis Boutan, who accomplished this off the Mediterranean coast of France in 1893. Six years later he used artificial light in underwater photography for the first time. Not only this, but the photographs were made at a depth of 165 feet. The Americans got on the bandwagon that year, when Simon Lake took photographs through a porthole of his submerged *Argonaut.* Then in 1900, Boutan published a book descrbing the principles of underwater photography. In a matter of a few years, there were several accomplished underwater photographers. In 1923, in the Florida Keys, W. H. Longley took the first underwater color photographs, and four years later he took the first artificially lighted colored photographs, some of which were published in the *National Geographic* the next year.

In 1903 Charles Williamson received a U.S. patent for a large flexible tube which telescoped down into the sea, reportedly to a depth of 300 feet. It was intended that a person climb down a ladder-like interior to work from a chamber at the bottom of an air-filled tube. Several pairs of flexible arms were described as projecting from the sides of the chamber, to permit men to work on sunken ships. After a number of unsuccessful tests it was determined that the device wasn't capable of working at such depths. By 1913 the inventor's son, J. E. Williamson, changed the work chamber into a "photosphere" by installing a thick glass window on one side. Through the window, Williamson could take both still and motion

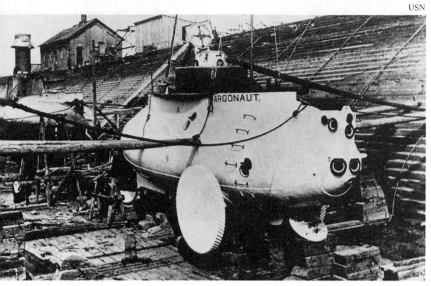

Simon Lake's submarine Argonaut *under construction at Baltimore, Md., circa 1898.*

Piccard's Ben Franklin *submersible.*

LEFT: *Dimitri Rebikoff's one-man submarine* Pegasus.
RIGHT: *Divers from Jacques-Yves Cousteau's underwater colony in the Red Sea collect specimens of fish in plastic bags.*

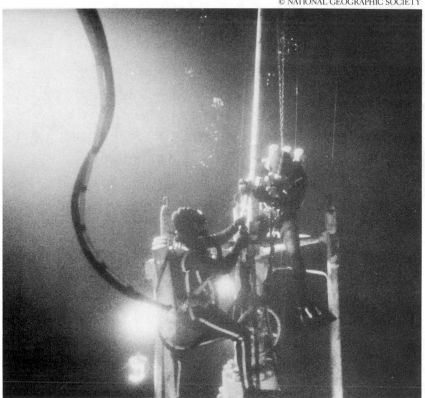

French oceanauts working on an oil-well head, 370 feet deep in the Mediterranean.

pictures. He showed them to a large group, on the strength of which Williamson obtained financial backing in 1915 to film Jules Verne's *20,000 Leagues Under the Sea*. The filming was done at an average depth of 30 feet in the clear waters of the Caribbean, and the movie was a great success. Williamson had this to say about the use of SCUBA divers in the movie.

The divers were equipped with self-contained diving suits, which are diving suits that have no connection with the surface. The air supply is renewed and purified by a chemical known as oxylite, carried in the diver's air containers. We had been running short of this chemical and the men were forced to use old charges; a dangerous practice, for when their one hour of usefulness was up, the foul gas coming from it suddenly intoxicated a diver so that, before he realized it, he was a drunken maniac. Even after we replenished the supply of this chemical and progressed with our undersea work, I discovered that the divers, stimulated by the effects of the chemicals they breathed, felt dreamily happy after being down three or four hours. Often, after playing their parts in the scene, they would wander away on excursions of their own, exploring coral caves or picking anemones and would be missing when we required them for the next scene. It was a curious experience to have these hardened veterans of serious work go off picking flowers like children. It was also dangerous. On one occasion, working overtime in an emergency, we came uncomfortably close to tragedy. The divers had emerged from the air-lock of the *Nautilus* to the seafloor, and had been working a long time in a particularly [important] scene. As they prepared to reenter the submarine, they suddenly fell upon one another like maniacs, fighting desperately. Struggling, utterly crazed by the exhausted chemical, one of the men was caught in the current, swept off his feet, and knocking a valve which inflated his suit, was sent thrashing aimlessly towards the surface, where an alert native diver rescued him.

From this description one gets the idea that making underwater movies in those days was far from safe or simple. But it was exciting and profitable, and soon others were entering the field. In 1931 E. R. F. Johnson introduced the first commercial 16-millimeter underwater motion picture camera and two years later, designed a housing for a deep-water movie camera which, unfortunately, was lost in a test at 18,000 feet. In 1939 a remote motion picture camera was successfully tested at a depth of 4,200 feet by two Americans, E. R. Baylor and E. N. Harvey. Two years earlier, Yves le Prieur had first used an underwater movie camera for military purposes, by filming a submarine escape exercise off the coast of France. The first outstanding underwater movies of marine life were made by Hans Hass in the West Indies, and the following year Cousteau began filming the first of his many exciting undersea films.

The first successful underwater robot camera for deep-sea time-lapse photography was introduced in 1913 and tested off Monaco. Three years later the camera, with floodlights added, was used to locate German booby traps in flooded Belgian coal mines. Also that year, the U.S. Navy began testing it for possible military use, and soon thereafter oceanographers used it in their research. In 1941 American oceanographer Maurice Ewing obtained some startling photographs at a depth of 16,200 feet, and within a decade, the robot camera was being used in the most extreme ocean depths. At M.I.T. Harold E. Edgerton made significant improvements on the underwater still and motion picture cameras, in addition to inventing strobe lights, which were used on all of the deep dives made by the *Trieste* and other research submarines. His equipment is still being used in oceanographic deep-water photography. In addition to designing the strobe robot camera, Edgerton also developed the electronic flash—which is used both topside and underwater—and several types of underwater sonar units for oceanographic, commercial, and military uses. Since the late 1950s Edgerton has been participating in underwater archeological expeditions in various parts of the world. Most of them were successful because of his equipment and expertise. In 1968 his sonar was instrumental in finding two of Columbus's shipwrecks which were lost off Jamaica in 1504, and in 1974 it located, and his deepsea cameras identified, the hulk of the Civil War warship, *Monitor*, lost off Cape Hatteras.

In 1925 H. Hartman, who was in charge of the navy's experimental robot depth-cameras, supervised the first tests of television underwater off the Isle of Capri. The first major use of underwater television cameras was in connection with the A-bomb tests in 1947 when scientists wanted to determine the amount of damage done to the ships sunk by the explosion, and it was too dangerous to send divers down in the radioactive waters. Four years later the camera proved itself again. The British Admiralty was desperately searching the Thames estuary for the lost submarine *Affray*, and the searchers had located some potential targets. A television camera used to identify the targets finally located the missing submarine. No one could understand why the crew hadn't tried to escape. One theory was that they had all been killed instantly when the submarine plunged to the bottom. The British again used underwater television in 1954 when a BOAC Comet airliner exploded in midair off the island of Elba and crashed in the sea, whereupon it sank to a depth of 430 feet. After a number of objects were located by a wire-sweep operation, the TV camera was able to identify fragments of the plane and to assist in their recovery by a four-ton grab bucket. During the search the camera accidentally found the remains of an ancient Greek sailing vessel. Among the many amphorae seen was a four-foot statue of a nude female.

"Genius" is a word often used to describe moderately brilliant people. But it aptly describes Dimitri Rebikoff, a pioneer in underwater photography and manned vehicles, as well as underwater archeology in the Mediterranean. In 1946 he developed the first portable strobe unit, which paralleled Dr. Edgerton's work at M.I.T. Next, he invented the first fully automatic exposure meter for cameras and designed the first high-powered microsecond strobe for color ballistics studies. Rebikoff got his first taste of diving and underwater photography in 1948 under the watchful eye of the famous American photographer, Phillipe Halsman, in the frigid waters of Lake Geneva in Switzerland. He then moved to the warmer Mediterranean where he became interested in underwater photography. To record the archeological sites he was exploring, he perfected his strobe unit for use underwater. The new unit, which he called "the Torpedo," consisted of a high-voltage battery-powered strobe light housed in a long Plexiglas tube filled with clear transformer oil. It was installed forward of the camera lens to avoid lighting up the particles which were in suspension between the subject and lens. Rebikoff improved on the Torpedo, converting it into one of the earliest underwater scooters by placing a propeller on the rear and using an electric engine to power it. Discovering that the normal lens of an underwater camera not only increased the focal length by 34 percent but also caused optical aberrations such as sphericity, chromatism, and astigmatism, Rebikoff and another scientist developed the Ivanhoff-Rebikoff Corrected Lens which corrected these aberrations.

To photogrammetically map ancient shipwrecks and obtain plani-metric and elevation measurements, Rebikoff, in 1954, developed a stereo underwater camera with a corrected lens. A wreck site which normally took several weeks for a team of divers to map could now be mapped in an hour. That year he also invented and built the *Pegasus,* the first one-man submarine capable of carrying a diver, photographic equipment, and other instrumentation to a depth of 300 feet (later models went down 600 feet). Powered by a reliable electric motor with a reduction gear and controlled by an aircraft joystick, the sub gives excellent underwater maneuverability. It can turn in its own length (nine feet), go sideways, up, down, and can be stopped—all as fast as the diver can manipulate the controls. The *Pegasus* performs all the acrobatics of an airplane, without the danger of stalling, since it isn't as subject to gravity. Its speed is over three knots and it has greater endurance with a given amount of air, due to the fact that the consumption rate of air is roughly halved when a diver isn't expending energy by swimming. It increases the diver's range by a factor of 10, as well as providing the opportunity to photograph large underwater areas quickly and at a constant depth which is maintained by a navigation and control module on the unit.

During its first year of use by Rebikoff and the members of the Submarine Alpine Club of Cannes, France, more than 50 ancient shipwrecks were discovered off the coast of France and Italy in the Mediterranean. Over the years, fishermen had reported finding am-phorae and other objects from shipwrecks, in waters that were too deep for SCUBA divers or the *Pegasus.* Rebikoff solved the problem by devising a new unit, the "Poodle." He added three servo-control systems for pitch, yaw, and roll, and the unit was maneuvered by remote control from the surface through a tether cable. The diver was replaced by a wide-angle corrected-lens TV camera, forward-looking sonar, depth sounder, artificial horizon, and gyrocompass. During the first day of trials, the "Poodle" located two Phoenician shipwrecks, one at 540 feet and the other at 700, which showed beyond doubt that it could perform as well as the *Pegasus* but at much greater depths. It could also be used 24 hours a day, unre-stricted by the time a diver could spend underwater. Later improve-ments included more powerful floodlights, still and motion picture cameras, and grab-arms for retrieving items from shipwrecks.

"When once [the deep ocean] has been seen, it will remain forever the most vivid memory in life, solely because of its cosmic chill and isolation, the eternal and absolute darkness and the indescribable beauty of its inhabitants." So wrote naturalist William Beebe in 1934, describing the events leading up to his record-breaking dive to 3,028 feet inside his bathysphere. Before he and his engineer-associate Otis Barton descended to this unprecedented depth, man's penetration of the depths was less than 600 feet, and that only by diving in armored diving suits. Their bathysphere was a two-and-a-half ton steel ball 54 inches in interior diameter and an inch and a half thick, with a 14-inch entry hatch. They could see the marvels of the sea through three portholes and breathed pure oxygen at atmospheric pressure.

In spite of its success, however, the bathysphere had several deficiencies, in the view of the underwater biologist. Because of the cramped quarters, each dive was restricted to three and a half hours. Also, it would sink like a rock if the cable holding it ever broke; every up-and-down motion of the surface ship was transmitted to the sphere; and the floodlights were inadequate for illuminating the area surrounding the bathysphere. Nevertheless, Beebe and Barton achieved great success. During the 1930s and 40s they were as familiar to people interested in undersea activities as Cousteau is today.

In July 1969, while the world was engrossed in watching the Apollo moon flights, another important scientific expedition was taking place. This one, a voyage conceived and led by Jacques Piccard, was in the Gulf Stream. Although he has been to the deepest parts of the ocean and has made more than 150 deep dives, Piccard is fond of saying, "the most dangerous part of any voyage is the trip by car from home to port." His father, Auguste, was intrigued with deep-sea exploration, especially the exploits of Beebe and Barton in their bathysphere; but he regarded their means of reaching the ocean depths as primitive and inadequate. Piccard felt that they were foolish to trust their lives to one wire cable. He envisioned a heavy chamber supported by a buoyant float which could be made to sink by adding ballast and which could return to the surface by jettisoning the ballast.

World War II forced Auguste Piccard to drop his project temporarily but as soon as it was over, he was back at work. The bathyscaphe, the *FNRS-2*, was taken to the Cape Verde Islands in September 1948. His son, Jacques, was there for the tests. Several major problems developed, however, which meant that the tests were inconclusive. Surrendering to the outcries of the press that the tests were a failure and a waste of money, the old man turned the bathyscaphe over to the French navy. The Frenchmen made a few basic modifications, renaming the vessel, *FNRS-3*. Then, off Dakar on February 15, 1954, with George Houot and Pierre Willm aboard, a descent to a record depth of 13,282 feet was made.

Jacques was furious when he heard that the French were taking full credit for the dive. He vowed that he would build another bathyscaphe and continue his father's work. After raising the necessary capital, he built the *Trieste* in only 14 months. In design and construction, she was almost identical to the *FNRS-2*, except that she had a stronger float, which made her capable of resisting greater pressures. After several dozen successful dives in the Mediterranean, the bathyscaphe was turned over to oceanographers.

As early as 1951, when the British oceanographic vessel *Challenger II* discovered a hole 36,000 feet deep in the Pacific's Marianna Trench, young Piccard had dreamed of diving there. By spring 1955 the cost of operating the *Trieste* was becoming excessive, and he was having difficulty making ends meet. Then, after two years of not being able to raise money from oceanographic foundations, he was awarded a contract by the U.S. Office of Naval Research to conduct deep dives in the Mediterranean. The navy was so pleased with the tests that summer that it purchased the *Trieste* and hired Piccard as a consultant. More tests were made off southern California, and finally Jacques convinced the navy to attempt a deep dive.

At 8:00 A.M. on the morning of January 23, 1960—right on schedule, despite the fact that the weather was very bad and large seas were running—the *Trieste* began her descent, with Jacques Piccard and navy Lieutenant Don Walsh aboard. Everything went fine until they reached 32,500 feet, at which point they heard a dull cracking sound that made the cabin tremble. What they didn't know was that the Plexiglas viewing port on the entry tower had cracked. But somehow it held. At 1:06 P.M. they reached the bottom—35,800 feet—the deepest men had ever been, and the deepest they will ever go unless a deeper spot is found. They had settled on the bottom just a few yards from a fish of the sole family. This answered a question that had been puzzling scientists for years. It told them that fish could live even in the deepest ocean depths. Piccard and Walsh also saw some shrimp, probably attracted by the bright lights of the bathyscaphe. They radioed the surface vessel that they had reached the bottom and were carrying out scientific experiments. Then they began their ascent, reaching the surface at 4:56 that afternoon, four minutes ahead of schedule.

A few weeks later the two heroes were flown to Washington where they were given awards by President Eisenhower. The navy continued to use the *Trieste* for other projects. When the wreck of the nuclear submarine, *Thresher* was found, the bathyscaphe went down to photograph her. In her time the *Trieste* had made 128 deep dives. The navy later had her rebuilt, and she was renamed *Trieste II.*

Shortly after receiving the award in Washington, Piccard returned to Switzerland and began designing the world's first mesoscaphe, a middle-depth vehicle, for the Swiss exposition at the World's Fair. Named the *Auguste Piccard,* after Jacques' father who had died the previous year, it was launched in 1963. The vehicle was 93 feet long, large enough to hold 40 passengers with individual viewing ports. She closely resembled an airliner. The mesoscaphe was powered by electrically driven propellers and could move in any direction. It was capable of descending to a depth of 3,500 feet. Announced in the press as the "world's first tourist submarine," she made a total of 1,300 dives carrying some 33,000 passengers to the bottom of Lake Geneva, and ended up being used for offshore oil exploration in the Arabian Gulf.

In September 1964 Piccard announced that he wanted to build another mesoscaphe for the purpose of exploring the Gulf Stream off the east coast of the United States. He planned to use only the current for propulsion. His mesoscaphe would drift like an underwater *Kon Tiki.* By being completely silent and apparently motionless to the fish, the mesoscaphe would not disturb the environment or frighten the sea life. It would be an ideal research platform from which to make observations and listen to the infinite variety of sea noises. Piccard spent many anxious months trying to raise the necessary money, until the Grumman Aircraft Corporation, a firm which wasn't even engaged in oceanographic work at the time, decided to support the project. To cover the enormous expenses involved in the project, Grumman invited other oceanography-oriented organizations to participate. It was at this juncture that the Naval Oceanographic Office linked up with them.

The diving sub was christened *Ben Franklin* but was better known as the *PX-15*. She was 50 feet long, weighed 130 tons, and her 10-foot-diameter steel hull was equipped with 25 viewing portholes. Her operating depth was 2,000 feet, but she was safe to 4,000 feet. While no similar craft could stay down for more than 24 hours at a time, the *PX-15* was capable of weeks or even months at the bottom. Four motor-driven propellers at the corners of the hull could be used for either ascent or descent. The interior was fully equipped with a laboratory and living facilities for a six-man crew. At the completion of her trials in Lake Geneva, she was shipped to West Palm Beach, Florida where she underwent six additional months of trials. Then engineers installed the scientific equipment to be carried on the "Drift Mission."

Like the astronauts, the crew had to undergo intense physiological and medical tests before and after the drift. Their sleeping, eating, and toilet habits were to be carefully controlled and monitored, as were their assigned tasks. The very duration of the mission was of great interest to NASA, which was planning manned flights lasting for months. Men had never been confined in a hostile environment without in some way being supplied from the outside. The *PX-15* was to drift at an average of two knots at depths varying from 300 to 2,000 feet for 1,500 miles. The entire mission was to last 30 days.

The Drift Mission got underway July 14, 1969, with Piccard and a crew of five other oceanographers, including a representative from the British Royal Navy. After being towed out into the Gulf Stream, the *PX-15* descended to 1,500 feet and leveled off. Compared to most of Piccard's previous underwater explorations, this one was uneventful, even monotonous. The oceanographers were surprised at the relatively small number of fish they saw during the long voyage. A few squid, sharks, tuna, and porpoise were sighted, and once a large swordfish attacked them but quickly gave up when it discovered it couldn't pierce the thick hull. During the drift, several surprising discoveries were made. Off the South Carolina coast, at a depth of 600 feet, they encountered huge underwater waves. And near Cape Hatteras the *PX-15* was caught in a powerful side eddy which forced her out of the Gulf Stream. They had to be towed back on course by a surface support vessel.

Thirty-one days later, after drifting one thousand five hundred nautical miles, the *PX-15* surfaced about 300 miles southeast of Halifax, Nova Scotia. The crew emerged in good health and spirits, and still talking to one another. Overall, the mission was considered a success; much oceanographic data had been gathered.

Jacques Piccard has plans to build another diving vehicle, one that will be useful for ecological investigations. He believes "the sea will die from pollution unless man learns to control his poisoning of it."

15

UNDERWATER HABITATS

Man's ability to work deeper and more effectively under the sea has improved rapidly in the past 15 years since Hannes Keller's ill-fated 1,000-foot dive off Catalina Island in 1962. The increased effort was spurred by two important problems: the resources potential of the world's oceans and seabeds, and the growing strategic importance of the undersea domain. Large sums have been spent by industry and various governments in developing new subsea equipment and diving techniques. In terms of the depths divers can reach, the time they can spend, and the work they can accomplish, progress has been spectacular. All this has been made possible by the development of revolutionary saturation-diving techniques. The duration of deep dives has jumped from minutes or a few hours to days and even weeks, while there has been a corresponding increase in the depths that can be attained. Divers have already penetrated past the 2,000-foot level in simulated tank test dives, a depth that is expected to be doubled before long.

At the time Keller made his breakthrough into the deeper depths, major areas of the continental shelfs were inaccessible to divers using compressed air or even those breathing helium-oxygen mixtures, because remaining below for any length of time necessitated long decompression periods which were not only costly but dangerous. Something had to be done to enable divers to reach greater depths and remain there longer. At first it was thought that the experiments of Hannes Keller would provide the answer. But his mixture of gases was eventually found to be impractical, both because some of the gases are very rare and because a large support team and a lot of money are needed to maintain one or two divers.

Since 1950 the Office of Naval Research has been experimenting, trying to solve the problem. In 1957 a member of its staff, a surgeon named George Bond, made a significant contribution to diving technology when he discovered that use of the helium-oxygen mixture for a prolonged period of time, usually less than 24 hours, causes the tissues of the diver's body to become saturated with the gases. When this saturation occurs, the required decompression time before surfacing is the same whether a diver remains at a given depth for 24 hours or a month. Should a diver wish to stay underwater for a month, all that's necessary is that he be housed in an environment where the pressure of the gases is the same as the outside water pressure. Bond tested this theory by having three men live for 12 days in a pressurized chamber that simulated water pressure at 200 feet. At the end of the experiment, the men decompressed as though they had been in a depth of 200 feet for 24 hours, rather than 12 days, and suffered no ill effects.

Interest in Bond's concept of saturation diving was demonstrated by Cousteau who conducted his first test, "Conshelf I," in September 1962, maintaining two divers at a depth of 34 feet in undersea housing for a week, breathing compressed air. The next year, in the Red Sea, he conducted "Conshelf II" by constructing the first underwater village. It consisted of several huts at a depth of 31 feet. Five divers breathed compressed air there for 30 days. Another hut, which was positioned at a depth of 85 feet, housed two divers for a week; they breathed a mixture of 75 percent helium and 25 percent compressed air. The divers spent most of the daylight hours in the water, filming a movie and collecting biological and geological specimens. Inside their huts they enjoyed many of the comforts people enjoyed on the surface, including good food and wine, showers, even television. Their only complaint was the lack of female companionship.

Since the inception in 1959 of the "Man in the Sea" project, by Dr. Bond, Edwin Link, an American inventor and undersea explorer has also been involved in the project. For years Link was interested in the possibility of deep diving, primarily for its underwater archeological potential. He designed and built a diving chamber known as the "Link Cylinder," for experimental work in saturation diving. Two weeks before Conshelf I began, Link arrived on the French

Riviera to conduct experiments. Despite the fact that he was 58 at the time, he wasn't a man to ask anyone else to make a dangerous dive until he had first tested the system and was sure it was safe. On his first dive off Villefranche, he forgot to completely close the hatch on the bottom of the chamber, and enough water got in to cause it to plunge to the bottom. He later admitted that he would have drowned if the accident had occurred in deeper water.

On August 28, 1962 Link set a world record by staying in the chamber for eight hours at 60 feet. After it was brought to the surface, he had to spend seven hours decompressing. He planned to make a two-day dive to 200 feet, but the navy doctors who were observing the tests told him that it was too risky at his age, and Link reluctantly agreed, saying: "I don't think I should hog all of the glory." He chose a Belgian, Robert Sténuit, for the dive.

At 11:50 A.M. on September 6, Sténuit began his descent, reaching a depth of 200 feet 10 minutes later. Hot meals were lowered to him on the bottom, and, using a Hookah breathing unit, he would swim out of the chamber and pick up the food containers. He made diving excursions outside the shelter to depths of 243 feet. After 26 hours, however, rough seas caused a small boat carrying a vital supply of helium tanks from the shore to the *Sea Diver* to capsize, forcing Link to call a halt to the dive. After the chamber had been taken aboard, Sténuit had to spend 67 hours inside the chamber, decompressing. His only complaint about the long stay inside the chamber was that he got cold because he hadn't worn clothing that was warm enough.

The chamber was used after Sténuit had completed his decompression to raise the small boat and helium cylinders from a depth of 240 feet. This test showed that a diver could have stayed down for weeks, since the helium-oxygen mixture had produced no ill effects during the saturation dive. Now Link was ready for the next step in enabling man to penetrate deeper into the sea where he would live and work for long periods of time.

Link had originally planned to work with Cousteau in the next phase of the project, which was to place two men on the bottom in an underwater habitat and use the diving chamber as an elevator to lower and raise the men. But they disagreed on so many aspects of the project that Cousteau decided to go ahead on his own.

While preparing for this next phase, Link received research assistance from the National Geographic Society, the University of Pennsylvania, and the U.S. Navy. Under the supervision of Dr. Bond and Professor Christian Lambert, experimental "dry dives" were conducted in a diving chamber at the Experimental Diving Unit in Washington, D.C., in which the depth that a diver could live at during saturation diving was progressively increased.

In April 1963 Link was preparing to sail from Monaco to Bermuda where he hoped to continue his deep-diving experiments in warmer waters, when he was urgently summoned to Washington. The *Thresher* had sunk mysteriously, and the navy was hastily forming the Deep Submergence System Research Group, which Link was invited to join. Nuclear-powered subs operated in much deeper water than the earlier conventional submarines, and there had been no parallel development of new rescue and salvage techniques. After months of study, Link and his colleagues recommended that future submarines be equipped with electronic devices that would enable search vessels to find them quickly. Also, a distress buoy should be carried on them, which would surface quickly in case of an accident and generate an S.O.S. signal. They further recommended that a manned undersea vehicle be developed which could do rescue work at depths up to 6,000 feet, yet be compact enough so it could be flown anywhere in the world in less than 24 hours. That vehicle, known as the DSRV, is now under construction.

Between meetings of the DSSRG, Link had designed a submersible, portable, inflatable dwelling (SPID), which was encased in rubber and would replace the diving chamber for future deep-diving experiments. When Sténuit saw the rubber casing on the SPID, he was skeptical about the walls' vulnerability to puncture. His doubts were well founded. One day during construction, a glass bottle of milk was left inside the dwelling. The milk fermented, causing the bottle to explode, and flying glass pierced the rubber walls of the dwelling. The rubber walls were reinforced with nylon.

For his next experiment Link needed a two-man team. One man wouldn't have been able to stay awake and maintain a constant watch over the instruments inside the SPID during the long period he would be on the bottom. Jon Lindbergh, son of the famous aviator and a man of considerable diving experience, was selected to make the dive with Sténuit. The goal was for the divers to live and work out of the SPID at a depth of about 400 feet for at least 48 hours. Under the medical supervision of Dr. Joseph Maginnis, there were numerous preliminary tests off Key West, at Miami, and in the Bahamas, before Link was satisfied.

For the experiment, he chose the Berry Islands in the Bahamas. The SPID was lowered and anchored on a sandy bottom 432 feet down. After extensive medical examinations, Sténuit and Lindbergh entered the diving chamber on the morning of June 30, 1964 and began their descent. When they reached the bottom they opened the hatch of the chamber and swam to the SPID. About seven feet long and four feet in diameter, it resembled a tent. The experiment went badly from the start. All instruments had been disconnected inside of the SPID before it was lowered, in case it accidentally flooded on the way down. So the first step was to reconnect the instruments and make sure they were functioning properly. The lights burned a few seconds and went out. Then a sealbeam light bulb exploded, spraying the interior with glass fragments. And the heat radiator wouldn't work. Using a flashlight for light, Stenuit attempted to start the carbon dioxide filtering machine, but it wouldn't work either. With no light, heat, or means of purifying their air, they, nevertheless, decided to stay down for at least 24 hours and take their chances. They signaled the *Sea Diver* about their plight, and Link signaled back that he was lowering a line, that they should send the defective filter unit to the surface. Four hours later it was sent back and to their relief, worked well. Without heat, though, it was a chilly night.

They spent most of the next day swimming around on the bottom, breathing through Hookah units connected to the SPID, which, they found, was the best way to keep warm at that depth. That evening they tackled the heat radiator and lights and managed to get them working properly. They spent the second night on the bottom in more comfortable circumstances than the first. With the exception of frequent interruptions caused by a large friendly grouper which kept bumping against the SPID, they had a good night's sleep.

After a hearty breakfast they spent the morning outside the SPID, exploring and playing with the inquisitive grouper. While they were eating lunch, Link sent down word that they had already spent over 48 hours on the bottom, and ordered them to surface. The divers would have gladly stayed down longer, now that all their problems were solved. Surfacing inside of the diving chamber, which was placed aboard the *Sea Diver* and then connected to a roomier and more comfortable deck decompression chamber, they spent 96 hours decompressing to normal atmospheric conditions. Actually, the decompression period would not have been longer even if they had been down for months, since their blood's nitrogen-saturation level had been reached.

The experiment was a success. It was the longest dive made at such a depth. Hannes Keller had gone deeper, but for only a shorter period. The next month the navy conducted its Sea Lab I experiment off Bermuda, in which four divers remained on the bottom for 11 days, but only at a depth of 192 feet, and at a high monetary cost. Link's "Man in the Sea" experiment had not only established decompression schedules for deep-saturation dives, all of which had previously been conducted only in laboratories, but had also proved the feasibility of man living and working at these hitherto unattained depths. Link also demonstrated that a small qualified team could work at these depths at a reasonable cost and with limited support, unlike the navy's elaborate Sea Lab projects.

In Sea Lab I, four navy divers, or "aquanauts," lived in a steel chamber 40 feet long and 10 feet wide. It contained sleeping quarters, galley, shower, toilets, work space, and a storeroom. There were four large windows through which the aquanauts could observe. In addition, there were communications with the surface support ship. During their stay the aquanauts followed a schedule designed to prove that man can work as well in the depths of the sea as on land, if the proper conditions are provided. The men spent hours each day observing and making sound recordings of marine creatures, collecting geological and biological samples, taking photographs, surveying the seafloor, and experimenting with power tools.

The results were so encouraging that the navy followed it with Sea Lab II, by far the longest and most ambitious deep-sea diving project ever conducted. It took place off La Jolla, California from August 28 to October 12, 1965. The undersea dwelling, located in 205 feet of water, was larger and better equipped than the one used in Sea Lab I. During the 45 days that Sea Lab II lasted, Captain George Bond maintained constant surveillance of the aquanauts by means of television and telephone, gathering a great amount of physiological data on living and working underwater.

Thirty divers, divided into teams of 10 each, were used. The first team was led by Commander Scott Carpenter, an astronaut turned aquanaut. The divers in this group were to get the dwelling operating properly, explore and map the surrounding seafloor, set up an underwater weather station, make marine biological studies, and perform human physiological experiments. It was a busy schedule for anyone, particularly divers trying to adjust to a new environment with the aid of equipment that was, for the most part, being used for the first time. Over 15 days they spent 6,067 minutes in the water, performing their tasks. On September 12 the first team—with the exception of Carpenter, who remained below for another 15 days—returned to the surface ship in a diving chamber, undergoing only 31 hours of decompression in a special chamber.

The second team, also commanded by Carpenter, entered the dwelling shortly after the first team left. Five members of the team were oceanographers. Their principal job was to study marine life. One of their most interesting experiments involved a porpoise named Tuffy. The experiment proved the value of these highly intelligent mammals to mankind. Tuffy was an expert mailman, delivering packages to and from the surface. He was also a friendly guide, leading aquanauts who pretended to be lost back to their dwelling. A major achievement of the second team was an excursion to a depth of 266 feet. In 15 days they spent 5,850 minutes in the water. On September 27 all were brought up to the surface, including Carpenter this time, where everyone underwent 35 hours of decompression—four hours longer than the first team because of the greater depth of their dives.

The third team, led by Master Navy Diver Robert Sheats, entered the dwelling minutes after the second team surfaced. The primary job of these aquanauts was to experiment with new salvage procedures and mining techniques. They tested a special foam for salvaging an airplane that had been deliberately sunk near the dwelling, as well as testing power and explosive tools designed to attach plates and lifting gear to sunken wrecks. In their mining experiments they tested vacuum devices for mining minerals on the seafloor, in addition to coring tubes and other devices for collecting mineral samples. During their stay they made dives to 300 feet where they remained for considerable periods of time. Before surfacing, they prepared the dwelling for raising. All aspects of the experiment were a success.

The U.S. Navy's underwater habitat Sealab II.

The Aluminaut *submersible.*

The U.S. Navy's submersible Sea-Craft *being hoisted ashore.*

The submersible Alvin *on its 1966 mission to find a missing American H-bomb off the coast of Spain.*

There were no casualties or injuries, and knowledge of living and working at great depths was substantially increased.

Concurrent with Sea Lab II was an even more adventurous experiment—Conshelf III. Captain Cousteau commanded the project, which was over a hundred feet deeper than American aquanauts had gone. He constructed a spherical undersea dwelling 18 feet in diameter, which was divided into two stories. The lower story contained sleeping quarters, sanitation facilities, and a storeroom for diving equipment, while the upper story contained dining and living quarters and the communications equipment. Seventy-seven tons of ballast were required to sink the dwelling, and the surface support consisted of a dozen vessels and 150 technicians. The six men who were to live underwater trained intensively for over a year.

On September 17, 1965 the men entered the dwelling on the surface of Monaco Harbor and laid on 11 atmospheres of heliox (a breathing mixture 98 percent helium and 2 percent oxygen), or a pressure equivalent to that of water at 328 feet. The dwelling was towed seven miles to the Cap Ferrat Lighthouse which was to serve as both power center and Cousteau's command post. The descent was delayed for four days when a storm arose, which lashed the dwelling and tow vessels and snapped some of the power and communications cables. The dwelling was towed back to the calmer waters of Monaco Harbor, with the divers inside, where they waited for better weather and cable repairs to be made. On the morning of September 21, the dwelling was once again towed to the lighthouse, and shortly after midnight was sunk to a depth of 328 feet. The experiment had begun.

Living conditions were much the same as those in Sea Lab II. The divers in both projects could compare notes by telephone even though they were separated by 6,000 miles. The tasks performed by the French resembled those of their American counterparts. In an important area they deserve credit for breaking new ground—the tapping of offshore oil deposits. Up to then, the high cost of offshore oil-drilling stations, or rigs, had made the recovery of oil from many deposits unprofitable. Cousteau's divers pioneered a new method. A partially erected oil derrick was lowered to them at a depth of 370 feet. They set it up on a slope 150 feet from the undersea dwelling, mounting the five-ton oil wellhead in only 45 minutes. During the remainder of their stay, they conducted experiments in running the well, demonstrating that oil can be recovered from deposits on the continental shelfs at a cost that is far less than any method previously known.

The success of the oil recovery and the demonstration that men can live and work in depths greater than anyone else had yet explored, made Conshelf III a noteworthy achievement. After three weeks on the bottom the aquanauts jettisoned the ballast, and the undersea dwelling rose to the surface where it was towed to Monaco Harbor. For another 88 hours, the required decompression time for them, the aquanauts remained inside the dwelling. On October 17 they emerged to breathe fresh air and receive worldwide acclaim.

To penetrate even deeper in the sea, Link developed a submersible vehicle called "Deep Diver," which can operate at a depth of 1,350 feet. It is constructed of two separate chambers. One compartment, kept at atmospheric pressure, houses the pilot, nondiving scientists, and technicians who, by watching through portholes or on closed-circuit television, can supervise and direct the work of divers. The other compartment, maintained at an atmospheric pressure equivalent to that of the surrounding water, carries the divers; it has a lockout exit to permit them to enter the water. During March 1968 Link had two of his divers—Roger Cook and Dennison Breese —look out from the Deep Diver at a depth of 705 feet, breathing mixed gases. They spent 25 minutes at this depth, conducting experiments under the watchful eyes of Dr. Joseph MacInnis who was acting as chief medical officer and copilot on the record-breaking dive.

The Sea Lab III project was to be the navy's most ambitious undertaking thus far. Three teams of 10 divers each were to live in a larger and more sophisticated habitat for several weeks at a depth of 610 feet on the ocean floor off San Clemente Island, California. There they would make sorties to a maximum depth of 1,000 feet, in addition to conducting scientific experiments. The project had already been delayed several times and was two years behind schedule when it finally got underway on February 15, 1969. After the habitat was lowered, instruments aboard the support ship indicated a helium leak and a consequent imbalance of the internal atmosphere of the habitat. Two days later, four divers were sent down in the personnel-transfer chamber to repair the leak. Diver Berry Cannon and his codivers were finning toward the habitat when TV monitors revealed that Cannon was swimming awkwardly and in obvious difficulty. Although he was immediately returned to the chamber and received heart massage and resucitation, Cannon died of heart failure. An autopsy showed that he had died of cardiorespiratory failure caused by carbon dioxide poisoning. How such a lethal quantity of the gas got into his breathing mixture was never revealed by the navy, and the entire project, which had already cost millions of dollars, was terminated.

After this, interest in habitats switched from trying to learn maximum capabilities to exploiting proven technology at relatively shallow depths. There has also been a switch from commercial and military applications to scientific study. At the shallower depths, costs are reduced, as well as risks, and nonprofessional divers can participate. Shortly after the Sea Lab III project was terminated, Tektite I was conducted. Four divers breathed mixed gases, living and working out of a habitat in 43 feet of water for 60 days. In 1970, the following year, Tektite II was conducted in the Virgin Islands. Eleven five-man teams (including women) spent between seven and 30 days each, for a total of 915 man-days, living in a habitat 40 feet down. Other experiments of this nature have been conducted in shallow waters.

The navy wasn't the only organization to suffer reverses. In October 1971 a pressurization failure in a diving bell resulted in the deaths of four Russian "oceanauts" during a simulated dive in a Moscow laboratory at a "depth" of 1,380 feet. The deep-diving research program was part of the Soviet Union's efforts to catch up in offshore oil exploration. For several years after Berry Cannon's death, the navy's "Man in the Sea" project was shelved, and many people feared that the navy would cancel the project permanently.

Then in 1972 the navy got back in the picture. Two navy divers, Warren Ramos Jr. and Chris Delucchi, made a record dive on mixed gases in the open ocean off San Clemente Island. Wearing heated suits and standard SCUBA equipment, with the special breathing mixture, they reached a depth of 945 feet where they remained for 30 minutes. Then they ascended and spent 10 days decompressing in a chamber. In June 1975 a team of U.S. Navy and British Royal Navy divers made a successful, open-sea dive to 1,148 feet in the Gulf of Mexico. After being taken down in a personnel-transfer chamber to a depth of 1,000 feet, the four divers went deeper on their own. One of them—William Rhodes—reached the bottom at 1,148 feet, the deepest open-sea dive yet made by a SCUBA diver.

When Sea Lab III was canceled, several oil firms began conducting deep-diving experiments on their own, and some achieved amazing results. In December 1970 the Taylor Diving and Salvage Company of Belle Chasse, Louisiana took five commercial divers on a 20-day simulated "wet" dive in a hyperbaric chamber to a depth of 1,100 feet. Although Taylor's divers had some minor difficulties, including a mild case of the bends, the chamber dive was rated as "successful" by navy observers. On June 19, 1973 three commercial British divers who were employed by Sub Sea Oil Services completed a 940-foot open-sea, working dive to inspect the foundation of an offshore drilling complex in the North Sea. Although they spent only 15 minutes at that depth, they required five days of decompression. Then during the summer of 1974, the French diving firm, COMEX, completed an experimental dive at its research facility in Marseille. Two divers were placed in a hyperbaric chamber for eight days at a pressure equivalent to 2,000 feet of water, and they suffered no ill effects.

Although a helium-oxygen mixture was used in COMEX's experimental dive, other scientists have found that it is unsafe to use such a mixture below 1,500 feet, because it becomes narcotic at that depth. The only other inert gas that can be used is hydrogen which, ironically, was the mixture Zetterstrom used in his deep-diving experiments in the 1940s. If the hydrogen-oxygen breathing gas can be developed for practical use, the physical properties of hydrogen might also lead to reduced decompression time for saturation diving, and man would be able to reach a depth of 5,000 feet. Beyond these depths some scientists believe it's possible, though not probable, that man can dive freely to 10,000 feet and deeper for short periods. When swimming at such great depths, the divers would breathe an oxygenated liquid pumped directly into their windpipes and lungs. Although the feasibility of man walking the abyssal plain is a matter for debate, his traditional urge to explore new worlds will unquestionably take him farther and farther into the depths of the sea.

The tragedy of the *Thresher* led to an intensive effort to develop small research submarines that can descend to great depths and perform rescue operations. Such submersible vehicles were badly needed for scientific exploration, as well as for the search for offshore oil deposits. These vehicles had a much better scientific-payload advantage over the larger bathyscaphes and surface-connected bathyspheres, enabling scientists to make fully integrated observations, using photographic, visual, acoustic, sampling, and measurement techniques, which are far less expensive. Likewise, they had a great advantage over the old method of obtaining data by lowering instruments by cable to great depths. Operating far below the often turbulent surface, these vehicles offer superior platform stability, as well as higher speeds at depth than do towed devices. In the five years following the *Thresher* disaster, funding from the DSSRG and private industry led to the development of dozens of these vehicles.

Actually, submersible vehicles had already been built, two of them in the 1950s. The *Submanaut* was originally designed to operate at a depth of 2,500 feet, but installation of a three-inch window for photography decreased this to 600 feet. Launched in 1956, the *Submanaut* was used to shoot underwater movies and was itself featured in the movie, *Around the World Under the Sea*. The other submersible vehicle—the *Goldfish*—was the creation of a former submariner. It was launched in 1958 and was used mainly for photographing shipwrecks underwater in settling insurance claims.

Displaying remarkable foresight in the requirements for manned submersibles, Cousteau as early as 1953 began designing a small vehicle capable of descending to 1,000 feet. His desire was to build a submersible from which scientists could observe and photograph undersea phenomena in comfort and with access to undersea valleys and narrow canyons, something that couldn't be done in the large, cumbersome bathyscaphes and bathyspheres. Cousteau had dived several times in the Piccard *FNRS-3*, and he knew the weak and strong points of the unit. He launched his first "Diving Saucer" in 1957, but it was lost shortly after that during an unmanned test dive. The lowering cable snapped and dropped it into 3,300 feet of water. In 1959 Cousteau built and launched his second Diving Saucer. At the sacrifice of great depth capability, he brought flexibility and range to underwater operations. He believed that speed was the enemy of observation, describing his slow-moving (0.6 knot cruising speed) saucer as "a scrutinizer, a loiterer, a deliberator, a taster of little scenes as well as big. She gave us six-hour periods in which to study accurately the things below up to a depth of 1,000 feet."

During its almost two decades of use the Diving Saucer has made thousands of safe deep-sea explorations.

Cousteau's saucer was the first genuine open-sea submersible vehicle system. Since its launching, more than 150 others have been built, most of which are used in offshore drilling for oil. By 1969 there were so many of them that, in order to demonstrate their capabilities, a jamboree of submersible vehicles was organized off Catalina Island. Each vehicle was designed for different tasks. The largest one was the cigar-shaped, 51-foot aluminum-hull *Aluminaut*, capable of four knots at a depth of 15,000, with an endurance of 32 hours. She carried a crew of four. Others were small by comparison. The *Star I* could carry only one man to a depth of 200 feet for four hours, and she had a cruising speed of one knot. The majority of the submersibles fell midway between these two examples—generally with a maximum operating depth of 5,000 feet and an endurance time of 10 to 12 hours, carrying a crew of two.

The usefulness of the submersibles was demonstrated when in 1966 an air force B-52 jet bomber collided with a jet tanker during refueling operations. Four H-bombs spilled out. Three of the bombs fell on land and were recovered, but the other one sank in the sea about six miles off the coast of Palomares, Spain, in about 2,550 feet of water. The search for the missing H-bomb began with U.S. Navy minesweeping ships. It soon became apparent that the problems of finding and recovering the bomb were beyond the capabilities of the minesweepers, so the navy called for outside help. Three diving subs were brought to the scene: the *Alvin*, good to 6,000 feet; the tiny *Cubmarine*, good only to 600 feet (at the time, no one had any idea how far offshore the bomb lay, or at what depth); and the *Aluminaut*. The navy also brought in its unmanned CURV (Cable-Controlled Underwater Research Vessel) recovery vehicle, which was a remote-controlled device designed to pick up practice torpedos from the ocean floor, down to depths of 2,000 feet. Normally, the CURV was lowered into the water from a surface ship, with control signals and electric power being sent down through a cable. TV pictures, showing the area in front of the unit, were sent back to the surface.

The search, which began on the day the bomb was lost (January 17, 1966), lasted until March 15 when it was found by the *Alvin*. Bad weather prevented recovery until March 24. Then the *Alvin* dived, and three lifting cables were attached to the bomb. *Alvin* could attach only one cable, however. When an attempt was made to lift the two-and-a-half-ton bomb, the cable snapped, and the bomb was lost again. Nine days of searching followed. The *Alvin* again found it, this time 300 feet deeper. Although the CURV recovery vehicle was designed to operate down to 2,000 feet, the navy hastily modified it to work deeper, and it was used to attach the cables to the bomb. During the attempt, it became entangled in the parachute that had been attached to the bomb. So, on April 7, the whole thing was raised.

Two years later, the *Alvin*, which was involved in oceanographic research off Cape Cod at the time, accidently sank when her lowering cables broke on the surface; she went down in over 5,000 feet of water. There was just enough air inside the craft to keep it afloat a few seconds, but this was enough for the three-man crew to escape. For awhile everyone thought the *Alvin* was lost for good. Then the navy authorized the money for her recovery. The *Aluminaut* was called in and performed the task, in much the same manner that the *Alvin* had been used to lift the H-bomb.

Plans for the future sound like science fiction: undersea cities being built; porpoise and dolphins being trained in undersea warfare; scientists claiming that membranes and gills can be attached to the human body, thus enabling people to breathe underwater; and mining companies preparing to harvest minerals from the deepest parts of the ocean. The race to penetrate the sea continues. It is as exciting and heroic as the space race. Many experts believe that before long, the regions of the earth lying underwater will be as well charted as the earth's land surface, and that the vast store of natural resources which lie at the bottom of the sea could eventually be the answer to the needs of an expanding world population.

SUGGESTED READING

Barton, Otis. *The World Beneath the Sea.* New York: Thomas Y. Crowell Co., 1953.

Bass, George F. *Archaeology Under Water.* New York: Frederick A. Praeger, Inc., 1966.

Beach, Edward L. *Submarine.* New York: Henry Holt and Company, Inc., 1952.

Beebe, William. *Half Mile Down.* New York: Harcourt, Brace & Company, 1939.

Blair, Clay, Jr. *The Atomic Submarine.* New York: Henry Holt and Company, Inc., 1954.

————. *Diving for Pleasure and Treasure.* Cleveland: The World Publishing Company, 1960.

Blassingame, Wyatt. *The U.S. Frogmen of World War II.* New York: Random House, Inc., 1964.

Borghese, Valerio. *The Sea Devils.* Chicago: Henry Regnery Company, 1954.

Burgess, Robert. *Ships Beneath the Sea.* New York, McGraw-Hill, 1975.

Busby, R. Frank. *Manned Submersibles.* Washington, D.C.: U.S. Navy, 1975.

Carson, Rachel. *The Sea Around Us.* New York: Oxford University Press, 1951.

Ciampi, Elgin. *Skin Diver.* New York: Ronald Press Co., 1960.

Clark, Eugenie. *Lady With a Spear.* New York: Harper & Brothers, 1953.

Cousteau, Jacques-Yves, and James Dugan. *The Living Sea.* New York: Harper & Row, 1963

Cousteau, Jacques-Yves, and Frédéric Dumas. *The Silent World.* New York: Harper & Brothers, 1953.

Davis, Robert H. *Deep Diving and Submarine Operations.* Rev. ed. Hackensack, N.J.: Wehman Bros., 1962.

De Borhegyi, Suzanne. *Ships, Shoals and Amphoras.* New York: Holt, Rinehart and Winston, Inc., 1961.

De Latil, Pierre, and Jean Rivoire. *Sunken Treasure.* New York: Hill & Wang, Inc., 1959.

————. *Man and the Underwater World.* New York: G.P. Putnam's Sons, 1956.

Doukan, Gilbert. *The World Beneath the Waves.* New York: John De Graff, Inc., 1957.

Dugan, James. *Man Under the Sea.* New York: Harper & Brothers, 1956.

Dugan, James, and Richard Vahan, eds. *Men Under Water.* New York: Chilton Co., 1965.

Ellsberg, Edward. *Men Under the Sea.* New York: Dodd, Mead & Company, Inc., 1939.

————. *On the Bottom.* New York: Dodd, Mead & Company, Inc., 1929.

Fane, Francis Douglas, and Don Moore. *The Naked Warriors.* New York: Appleton-Century-Crofts, Inc., 1956.

Franzen, Anders. *The Warship Vasa.* New York: W.S. Heinman, 1960.

Gores, Joseph N. *Marine Salvage.* New York: Doubleday & Co., 1971.

Hass, Hans. *Diving to Adventure.* New York: Doubleday and Company, Inc., 1951.

Houot, Georges, and Pierre Henri Willm. *2000 Fathoms Down.* New York: E.P. Dutton & Co., 1955.

Lake, Simon, and Herbert Corey. *Submarine.* New York: D. Appleton-Century Company, Inc., 1938.

Marx, Robert F. *Shipwrecks in the Americas.* New York: Dover Publications, Inc., 1987.

————. *The Underwater Dig.* New York: David McKay Co., 1975.

————. *The Lure of Sunken Treasure.* New York: David McKay Co. 1973.

————. *Port Royal Rediscovered.* New York: Doubleday & Co., 1973.

————. *Sea Fever.* New York: Doubleday & Co., 1972.

————. *Still More Adventures.* New York: Mason/Charter, 1976.

Masters, David. *Wonders of Salvage.* London: Eyre & Spottiswoode Ltd., 1944.

Owen, David M. *A Manual for Free Divers.* New York: Pergamon Press, Inc., 1955.

Potter, John S., Jr. *The Treasure Diver's Guide.* New York: Doubleday and Company, Inc., 1960.

Sténuit, Robert. *The Deepest Days.* New York: Coward-McCann, Inc., 1966.

Taylor, Joan du Plat. *Marine Archaeology.* New York: Thomas Y. Crowell Co., 1966.

ROBERT F. MARX is one of the most successful and well-known specialists in the field of marine archaeology and has an equally strong reputation for his work in naval and maritime history. His particular area of interest is Spanish naval activity in the Caribbean during the colonial period. He has written thirty-two books covering his wide range of interests and has published more than six hundred scientific and popular articles and reports. He studied at UCLA and at the University of Maryland and served in the United States Marine Corps, where he worked in diving and salvage operations.

He is perhaps best known to the general public for such feats as the voyage from Spain to San Salvador of the *Niña II*, a replica of Columbus's ship, on which he served as co-organizer and navigator. For this voyage he was made a Knight Commander in the Order of Isabella the Catholic by the Spanish government, the highest honorific order given to any foreigner. He also participated in the location of the Civil War ironclad *U.S.S. Monitor*, the discovery of several Mayan temple sites in Central America, the discovery of many famous Spanish, French, and English wrecks from the colonial period and, more recently, the mapping and excavation of the sunken city of Port Royal. He has also been an editor for *Argosy* and *The Saturday Evening Post*.

Robert F. Marx is shown above holding an amphora from a UNESCO-sponsored excavation of a sixth-century B.C. Phoenician wreck found in Tyre, Lebanon.

A CATALOG OF SELECTED
DOVER BOOKS
IN ALL FIELDS OF INTEREST

A CATALOG OF SELECTED DOVER
BOOKS IN ALL FIELDS OF INTEREST

DRAWINGS OF REMBRANDT, edited by Seymour Slive. Updated Lippmann, Hofstede de Groot edition, with definitive scholarly apparatus. All portraits, biblical sketches, landscapes, nudes. Oriental figures, classical studies, together with selection of work by followers. 550 illustrations. Total of 630pp. 9⅛ × 12¼.
21485-0, 21486-9 Pa., Two-vol. set $29.90

GHOST AND HORROR STORIES OF AMBROSE BIERCE, Ambrose Bierce. 24 tales vividly imagined, strangely prophetic, and decades ahead of their time in technical skill: "The Damned Thing," "An Inhabitant of Carcosa," "The Eyes of the Panther," "Moxon's Master," and 20 more. 199pp. 5⅜ × 8½. 20767-6 Pa. $3.95

ETHICAL WRITINGS OF MAIMONIDES, Maimonides. Most significant ethical works of great medieval sage, newly translated for utmost precision, readability. Laws Concerning Character Traits, Eight Chapters, more. 192pp. 5⅜ × 8½.
24522-5 Pa. $4.50

THE EXPLORATION OF THE COLORADO RIVER AND ITS CANYONS, J. W. Powell. Full text of Powell's 1,000-mile expedition down the fabled Colorado in 1869. Superb account of terrain, geology, vegetation, Indians, famine, mutiny, treacherous rapids, mighty canyons, during exploration of last unknown part of continental U.S. 400pp. 5⅜ × 8½. 20094-9 Pa. $7.95

HISTORY OF PHILOSOPHY, Julián Marías. Clearest one-volume history on the market. Every major philosopher and dozens of others, to Existentialism and later. 505pp. 5⅜ × 8½. 21739-6 Pa. $9.95

ALL ABOUT LIGHTNING, Martin A. Uman. Highly readable non-technical survey of nature and causes of lightning, thunderstorms, ball lightning, St. Elmo's Fire, much more. Illustrated. 192pp. 5⅜ × 8½. 25237-X Pa. $5.95

SAILING ALONE AROUND THE WORLD, Captain Joshua Slocum. First man to sail around the world, alone, in small boat. One of great feats of seamanship told in delightful manner. 67 illustrations. 294pp. 5⅜ × 8½. 20326-3 Pa. $4.95

LETTERS AND NOTES ON THE MANNERS, CUSTOMS AND CONDITIONS OF THE NORTH AMERICAN INDIANS, George Catlin. Classic account of life among Plains Indians: ceremonies, hunt, warfare, etc. 312 plates. 572pp. of text. 6⅛ × 9¼. 22118-0, 22119-9, Pa. Two-vol. set $17.90

ALASKA: The Harriman Expedition, 1899, John Burroughs, John Muir, et al. Informative, engrossing accounts of two-month, 9,000-mile expedition. Native peoples, wildlife, forests, geography, salmon industry, glaciers, more. Profusely illustrated. 240 black-and-white line drawings. 124 black-and-white photographs. 3 maps. Index. 576pp. 5⅜ × 8½. 25109-8 Pa. $11.95

THE BOOK OF BEASTS: Being a Translation from a Latin Bestiary of the Twelfth Century, T. H. White. Wonderful catalog real and fanciful beasts: manticore, griffin, phoenix, amphivius, jaculus, many more. White's witty erudite commentary on scientific, historical aspects. Fascinating glimpse of medieval mind. Illustrated. 296pp. 5⅜ × 8¼. (Available in U.S. only) 24609-4 Pa. $6.95

FRANK LLOYD WRIGHT: ARCHITECTURE AND NATURE With 160 Illustrations, Donald Hoffmann. Profusely illustrated study of influence of nature—especially prairie—on Wright's designs for Fallingwater, Robie House, Guggenheim Museum, other masterpieces. 96pp. 9¼ × 10¾. 25098-9 Pa. $7.95

FRANK LLOYD WRIGHT'S FALLINGWATER, Donald Hoffmann. Wright's famous waterfall house: planning and construction of organic idea. History of site, owners, Wright's personal involvement. Photographs of various stages of building. Preface by Edgar Kaufmann, Jr. 100 illustrations. 112pp. 9¼ × 10.
23671-4 Pa. $8.95

YEARS WITH FRANK LLOYD WRIGHT: Apprentice to Genius, Edgar Tafel. Insightful memoir by a former apprentice presents a revealing portrait of Wright the man, the inspired teacher, the greatest American architect. 372 black-and-white illustrations. Preface. Index. vi + 228pp. 8¼ × 11. 24801-1 Pa. $10.95

THE STORY OF KING ARTHUR AND HIS KNIGHTS, Howard Pyle. Enchanting version of King Arthur fable has delighted generations with imaginative narratives of exciting adventures and unforgettable illustrations by the author. 41 illustrations. xviii + 313pp. 6⅛ × 9¼. 21445-1 Pa. $6.95

THE GODS OF THE EGYPTIANS, E. A. Wallis Budge. Thorough coverage of numerous gods of ancient Egypt by foremost Egyptologist. Information on evolution of cults, rites and gods; the cult of Osiris; the Book of the Dead and its rites; the sacred animals and birds; Heaven and Hell; and more. 956pp. 6⅛ × 9¼.
22055-9, 22056-7 Pa., Two-vol. set $21.90

A THEOLOGICO-POLITICAL TREATISE, Benedict Spinoza. Also contains unfinished *Political Treatise*. Great classic on religious liberty, theory of government on common consent. R. Elwes translation. Total of 421pp. 5⅜ × 8½.
20249-6 Pa. $6.95

INCIDENTS OF TRAVEL IN CENTRAL AMERICA, CHIAPAS, AND YUCATAN, John L. Stephens. Almost single-handed discovery of Maya culture; exploration of ruined cities, monuments, temples; customs of Indians. 115 drawings. 892pp. 5⅜ × 8½. 22404-X, 22405-8 Pa., Two-vol. set $15.90

LOS CAPRICHOS, Francisco Goya. 80 plates of wild, grotesque monsters and caricatures. Prado manuscript included. 183pp. 6⅝ × 9⅝. 22384-1 Pa. $5.95

AUTOBIOGRAPHY: The Story of My Experiments with Truth, Mohandas K. Gandhi. Not hagiography, but Gandhi in his own words. Boyhood, legal studies, purification, the growth of the Satyagraha (nonviolent protest) movement. Critical, inspiring work of the man who freed India. 480pp. 5⅜ × 8½. (Available in U.S. only)
24593-4 Pa. $6.95

ILLUSTRATED DICTIONARY OF HISTORIC ARCHITECTURE, edited by Cyril M. Harris. Extraordinary compendium of clear, concise definitions for over 5,000 important architectural terms complemented by over 2,000 line drawings. Covers full spectrum of architecture from ancient ruins to 20th-century Modernism. Preface. 592pp. 7½ × 9⅝. 24444-X Pa. $15.95

THE NIGHT BEFORE CHRISTMAS, Clement Moore. Full text, and woodcuts from original 1848 book. Also critical, historical material. 19 illustrations. 40pp. 4⅝ × 6. 22797-9 Pa. $2.50

THE LESSON OF JAPANESE ARCHITECTURE: 165 Photographs, Jiro Harada. Memorable gallery of 165 photographs taken in the 1930's of exquisite Japanese homes of the well-to-do and historic buildings. 13 line diagrams. 192pp. 8⅜ × 11¼. 24778-3 Pa. $10.95

THE AUTOBIOGRAPHY OF CHARLES DARWIN AND SELECTED LETTERS, edited by Francis Darwin. The fascinating life of eccentric genius composed of an intimate memoir by Darwin (intended for his children); commentary by his son, Francis; hundreds of fragments from notebooks, journals, papers; and letters to and from Lyell, Hooker, Huxley, Wallace and Henslow. xi + 365pp. 5⅜ × 8. 20479-0 Pa. $6.95

WONDERS OF THE SKY: Observing Rainbows, Comets, Eclipses, the Stars and Other Phenomena, Fred Schaaf. Charming, easy-to-read poetic guide to all manner of celestial events visible to the naked eye. Mock suns, glories, Belt of Venus, more. Illustrated. 299pp. 5¼ × 8¼. 24402-4 Pa. $7.95

BURNHAM'S CELESTIAL HANDBOOK, Robert Burnham, Jr. Thorough guide to the stars beyond our solar system. Exhaustive treatment. Alphabetical by constellation: Andromeda to Cetus in Vol. 1; Chamaeleon to Orion in Vol. 2; and Pavo to Vulpecula in Vol. 3. Hundreds of illustrations. Index in Vol. 3. 2,000pp. 6⅛ × 9¼. 23567-X, 23568-8, 23673-0 Pa., Three-vol. set $38.85

STAR NAMES: Their Lore and Meaning, Richard Hinckley Allen. Fascinating history of names various cultures have given to constellations and literary and folkloristic uses that have been made of stars. Indexes to subjects. Arabic and Greek names. Biblical references. Bibliography. 563pp. 5⅜ × 8½. 21079-0 Pa. $8.95

THIRTY YEARS THAT SHOOK PHYSICS: The Story of Quantum Theory, George Gamow. Lucid, accessible introduction to influential theory of energy and matter. Careful explanations of Dirac's anti-particles, Bohr's model of the atom, much more. 12 plates. Numerous drawings. 240pp. 5⅜ × 8½. 24895-X Pa. $5.95

CHINESE DOMESTIC FURNITURE IN PHOTOGRAPHS AND MEASURED DRAWINGS, Gustav Ecke. A rare volume, now affordably priced for antique collectors, furniture buffs and art historians. Detailed review of styles ranging from early Shang to late Ming. Unabridged republication. 161 black-and-white drawings, photos. Total of 224pp. 8⅜ × 11¼. (Available in U.S. only) 25171-3 Pa. $13.95

VINCENT VAN GOGH: A Biography, Julius Meier-Graefe. Dynamic, penetrating study of artist's life, relationship with brother, Theo, painting techniques, travels, more. Readable, engrossing. 160pp. 5⅜ × 8½. (Available in U.S. only) 25253-1 Pa. $4.95

HOW TO WRITE, Gertrude Stein. Gertrude Stein claimed anyone could understand her unconventional writing—here are clues to help. Fascinating improvisations, language experiments, explanations illuminate Stein's craft and the art of writing. Total of 414pp. 4⅜ × 6⅜. 23144-5 Pa. $6.95

ADVENTURES AT SEA IN THE GREAT AGE OF SAIL: Five Firsthand Narratives, edited by Elliot Snow. Rare true accounts of exploration, whaling, shipwreck, fierce natives, trade, shipboard life, more. 33 illustrations. Introduction. 353pp. 5⅜ × 8½. 25177-2 Pa. $8.95

THE HERBAL OR GENERAL HISTORY OF PLANTS, John Gerard. Classic descriptions of about 2,850 plants—with over 2,700 illustrations—includes Latin and English names, physical descriptions, varieties, time and place of growth, more. 2,706 illustrations. xlv + 1,678pp. 8½ × 12¼. 23147-X Cloth. $75.00

DOROTHY AND THE WIZARD IN OZ, L. Frank Baum. Dorothy and the Wizard visit the center of the Earth, where people are vegetables, glass houses grow and Oz characters reappear. Classic sequel to *Wizard of Oz*. 256pp. 5⅜ × 8. 24714-7 Pa. $4.95

SONGS OF EXPERIENCE: Facsimile Reproduction with 26 Plates in Full Color, William Blake. This facsimile of Blake's original "Illuminated Book" reproduces 26 full-color plates from a rare 1826 edition. Includes "The Tyger," "London," "Holy Thursday," and other immortal poems. 26 color plates. Printed text of poems. 48pp. 5¼ × 7. 24636-1 Pa. $3.50

SONGS OF INNOCENCE, William Blake. The first and most popular of Blake's famous "Illuminated Books," in a facsimile edition reproducing all 31 brightly colored plates. Additional printed text of each poem. 64pp. 5¼ × 7. 22764-2 Pa. $3.50

PRECIOUS STONES, Max Bauer. Classic, thorough study of diamonds, rubies, emeralds, garnets, etc.: physical character, occurrence, properties, use, similar topics. 20 plates, 8 in color. 94 figures. 659pp. 6⅛ × 9¼. 21910-0, 21911-9 Pa., Two-vol. set $15.90

ENCYCLOPEDIA OF VICTORIAN NEEDLEWORK, S. F. A. Caulfeild and Blanche Saward. Full, precise descriptions of stitches, techniques for dozens of needlecrafts—most exhaustive reference of its kind. Over 800 figures. Total of 679pp. 8⅛ × 11. Two volumes. Vol. 1 22800-2 Pa. $11.95
Vol. 2 22801-0 Pa. $11.95

THE MARVELOUS LAND OF OZ, L. Frank Baum. Second Oz book, the Scarecrow and Tin Woodman are back with hero named Tip, Oz magic. 136 illustrations. 287pp. 5⅜ × 8½. 20692-0 Pa. $5.95

WILD FOWL DECOYS, Joel Barber. Basic book on the subject, by foremost authority and collector. Reveals history of decoy making and rigging, place in American culture, different kinds of decoys, how to make them, and how to use them. 140 plates. 156pp. 7⅞ × 10¾. 20011-6 Pa. $8.95

HISTORY OF LACE, Mrs. Bury Palliser. Definitive, profusely illustrated chronicle of lace from earliest times to late 19th century. Laces of Italy, Greece, England, France, Belgium, etc. Landmark of needlework scholarship. 266 illustrations. 672pp. 6⅛ × 9¼. 24742-2 Pa. $14.95

ILLUSTRATED GUIDE TO SHAKER FURNITURE, Robert Meader. All furniture and appurtenances, with much on unknown local styles. 235 photos. 146pp. 9 × 12. 22819-3 Pa. $8.95

WHALE SHIPS AND WHALING: A Pictorial Survey, George Francis Dow. Over 200 vintage engravings, drawings, photographs of barks, brigs, cutters, other vessels. Also harpoons, lances, whaling guns, many other artifacts. Comprehensive text by foremost authority. 207 black-and-white illustrations. 288pp. 6 × 9. 24808-9 Pa. $8.95

THE BERTRAMS, Anthony Trollope. Powerful portrayal of blind self-will and thwarted ambition includes one of Trollope's most heartrending love stories. 497pp. 5⅜ × 8½. 25119-5 Pa. $9.95

ADVENTURES WITH A HAND LENS, Richard Headstrom. Clearly written guide to observing and studying flowers and grasses, fish scales, moth and insect wings, egg cases, buds, feathers, seeds, leaf scars, moss, molds, ferns, common crystals, etc.—all with an ordinary, inexpensive magnifying glass. 209 exact line drawings aid in your discoveries. 220pp. 5⅜ × 8½. 23330-8 Pa. $4.95

RODIN ON ART AND ARTISTS, Auguste Rodin. Great sculptor's candid, wide-ranging comments on meaning of art; great artists; relation of sculpture to poetry, painting, music; philosophy of life, more. 76 superb black-and-white illustrations of Rodin's sculpture, drawings and prints. 119pp. 8⅜ × 11¼. 24487-3 Pa. $7.95

FIFTY CLASSIC FRENCH FILMS, 1912–1982: A Pictorial Record, Anthony Slide. Memorable stills from Grand Illusion, Beauty and the Beast, Hiroshima, Mon Amour, many more. Credits, plot synopses, reviews, etc. 160pp. 8¼ × 11. 25256-6 Pa. $11.95

THE PRINCIPLES OF PSYCHOLOGY, William James. Famous long course complete, unabridged. Stream of thought, time perception, memory, experimental methods; great work decades ahead of its time. 94 figures. 1,391pp. 5⅜ × 8½. 20381-6, 20382-4 Pa., Two-vol. set $23.90

BODIES IN A BOOKSHOP, R. T. Campbell. Challenging mystery of blackmail and murder with ingenious plot and superbly drawn characters. In the best tradition of British suspense fiction. 192pp. 5⅜ × 8½. 24720-1 Pa. $3.95

CALLAS: PORTRAIT OF A PRIMA DONNA, George Jellinek. Renowned commentator on the musical scene chronicles incredible career and life of the most controversial, fascinating, influential operatic personality of our time. 64 black-and-white photographs. 416pp. 5⅜ × 8¼. 25047-4 Pa. $8.95

GEOMETRY, RELATIVITY AND THE FOURTH DIMENSION, Rudolph Rucker. Exposition of fourth dimension, concepts of relativity as Flatland characters continue adventures. Popular, easily followed yet accurate, profound. 141 illustrations. 133pp. 5⅜ × 8½. 23400-2 Pa. $3.95

HOUSEHOLD STORIES BY THE BROTHERS GRIMM, with pictures by Walter Crane. 53 classic stories—Rumpelstiltskin, Rapunzel, Hansel and Gretel, the Fisherman and his Wife, Snow White, Tom Thumb, Sleeping Beauty, Cinderella, and so much more—lavishly illustrated with original 19th century drawings. 114 illustrations. x + 269pp. 5⅜ × 8½. 21080-4 Pa. $4.95

SUNDIALS, Albert Waugh. Far and away the best, most thorough coverage of ideas, mathematics concerned, types, construction, adjusting anywhere. Over 100 illustrations. 230pp. 5⅜ × 8½. 22947-5 Pa. $4.95

PICTURE HISTORY OF THE NORMANDIE: With 190 Illustrations, Frank O. Braynard. Full story of legendary French ocean liner: Art Deco interiors, design innovations, furnishings, celebrities, maiden voyage, tragic fire, much more. Extensive text. 144pp. 8⅞ × 11¾. 25257-4 Pa. $10.95

THE FIRST AMERICAN COOKBOOK: A Facsimile of "American Cookery," 1796, Amelia Simmons. Facsimile of the first American-written cookbook published in the United States contains authentic recipes for colonial favorites— pumpkin pudding, winter squash pudding, spruce beer, Indian slapjacks, and more. Introductory Essay and Glossary of colonial cooking terms. 80pp. 5⅜ × 8½. 24710-4 Pa. $3.50

101 PUZZLES IN THOUGHT AND LOGIC, C. R. Wylie, Jr. Solve murders and robberies, find out which fishermen are liars, how a blind man could possibly identify a color—purely by your own reasoning! 107pp. 5⅜ × 8½. 20367-0 Pa. $2.50

THE BOOK OF WORLD-FAMOUS MUSIC—CLASSICAL, POPULAR AND FOLK, James J. Fuld. Revised and enlarged republication of landmark work in musico-bibliography. Full information about nearly 1,000 songs and compositions including first lines of music and lyrics. New supplement. Index. 800pp. 5⅜ × 8¼. 24857-7 Pa. $15.95

ANTHROPOLOGY AND MODERN LIFE, Franz Boas. Great anthropologist's classic treatise on race and culture. Introduction by Ruth Bunzel. Only inexpensive paperback edition. 255pp. 5⅜ × 8½. 25245-0 Pa. $6.95

THE TALE OF PETER RABBIT, Beatrix Potter. The inimitable Peter's terrifying adventure in Mr. McGregor's garden, with all 27 wonderful, full-color Potter illustrations. 55pp. 4¼ × 5½. (Available in U.S. only) 22827-4 Pa. $1.75

THREE PROPHETIC SCIENCE FICTION NOVELS, H. G. Wells. *When the Sleeper Wakes, A Story of the Days to Come* and *The Time Machine* (full version). 335pp. 5⅜ × 8½. (Available in U.S. only) 20605-X Pa. $6.95

APICIUS COOKERY AND DINING IN IMPERIAL ROME, edited and translated by Joseph Dommers Vehling. Oldest known cookbook in existence offers readers a clear picture of what foods Romans ate, how they prepared them, etc. 49 illustrations. 301pp. 6⅛ × 9¼. 23563-7 Pa. $7.95

SHAKESPEARE LEXICON AND QUOTATION DICTIONARY, Alexander Schmidt. Full definitions, locations, shades of meaning of every word in plays and poems. More than 50,000 exact quotations. 1,485pp. 6½ × 9¼.
22726-X, 22727-8 Pa., Two-vol. set $29.90

THE WORLD'S GREAT SPEECHES, edited by Lewis Copeland and Lawrence W. Lamm. Vast collection of 278 speeches from Greeks to 1970. Powerful and effective models; unique look at history. 842pp. 5⅜ × 8½. 20468-5 Pa. $11.95

THE BLUE FAIRY BOOK, Andrew Lang. The first, most famous collection, with many familiar tales: Little Red Riding Hood, Aladdin and the Wonderful Lamp, Puss in Boots, Sleeping Beauty, Hansel and Gretel, Rumpelstiltskin; 37 in all. 138 illustrations. 390pp. 5⅜ × 8½.   21437-0 Pa. $6.95

THE STORY OF THE CHAMPIONS OF THE ROUND TABLE, Howard Pyle. Sir Launcelot, Sir Tristram and Sir Percival in spirited adventures of love and triumph retold in Pyle's inimitable style. 50 drawings, 31 full-page. xviii + 329pp. 6½ × 9¼.   21883-X Pa. $7.95

AUDUBON AND HIS JOURNALS, Maria Audubon. Unmatched two-volume portrait of the great artist, naturalist and author contains his journals, an excellent biography by his granddaughter, expert annotations by the noted ornithologist, Dr. Elliott Coues, and 37 superb illustrations. Total of 1,200pp. 5⅜ × 8.
Vol. I 25143-8 Pa. $8.95
Vol. II 25144-6 Pa. $8.95

GREAT DINOSAUR HUNTERS AND THEIR DISCOVERIES, Edwin H. Colbert. Fascinating, lavishly illustrated chronicle of dinosaur research, 1820's to 1960. Achievements of Cope, Marsh, Brown, Buckland, Mantell, Huxley, many others. 384pp. 5¼ × 8¼.   24701-5 Pa. $7.95

THE TASTEMAKERS, Russell Lynes. Informal, illustrated social history of American taste 1850's–1950's. First popularized categories Highbrow, Lowbrow, Middlebrow. 129 illustrations. New (1979) afterword. 384pp. 6 × 9.
23993-4 Pa. $8.95

DOUBLE CROSS PURPOSES, Ronald A. Knox. A treasure hunt in the Scottish Highlands, an old map, unidentified corpse, surprise discoveries keep reader guessing in this cleverly intricate tale of financial skullduggery. 2 black-and-white maps. 320pp. 5⅜ × 8½. (Available in U.S. only)   25032-6 Pa. $6.95

AUTHENTIC VICTORIAN DECORATION AND ORNAMENTATION IN FULL COLOR: 46 Plates from "Studies in Design," Christopher Dresser. Superb full-color lithographs reproduced from rare original portfolio of a major Victorian designer. 48pp. 9¼ × 12¼.   25083-0 Pa. $7.95

PRIMITIVE ART, Franz Boas. Remains the best text ever prepared on subject, thoroughly discussing Indian, African, Asian, Australian, and, especially, Northern American primitive art. Over 950 illustrations show ceramics, masks, totem poles, weapons, textiles, paintings, much more. 376pp. 5⅜ × 8. 20025-6 Pa. $6.95

SIDELIGHTS ON RELATIVITY, Albert Einstein. Unabridged republication of two lectures delivered by the great physicist in 1920–21. *Ether and Relativity* and *Geometry and Experience*. Elegant ideas in non-mathematical form, accessible to intelligent layman. vi + 56pp. 5⅜ × 8½.   24511-X Pa. $2.95

THE WIT AND HUMOR OF OSCAR WILDE, edited by Alvin Redman. More than 1,000 ripostes, paradoxes, wisecracks: Work is the curse of the drinking classes, I can resist everything except temptation, etc. 258pp. 5⅜ × 8½.   20602-5 Pa. $4.95

ADVENTURES WITH A MICROSCOPE, Richard Headstrom. 59 adventures with clothing fibers, protozoa, ferns and lichens, roots and leaves, much more. 142 illustrations. 232pp. 5⅜ × 8½.   23471-1 Pa. $3.95

PLANTS OF THE BIBLE, Harold N. Moldenke and Alma L. Moldenke. Standard reference to all 230 plants mentioned in Scriptures. Latin name, biblical reference, uses, modern identity, much more. Unsurpassed encyclopedic resource for scholars, botanists, nature lovers, students of Bible. Bibliography. Indexes. 123 black-and-white illustrations. 384pp. 6 × 9. 25069-5 Pa. $8.95

FAMOUS AMERICAN WOMEN: A Biographical Dictionary from Colonial Times to the Present, Robert McHenry, ed. From Pocahontas to Rosa Parks, 1,035 distinguished American women documented in separate biographical entries. Accurate, up-to-date data, numerous categories, spans 400 years. Indices. 493pp. 6½ × 9¼. 24523-3 Pa. $10.95

THE FABULOUS INTERIORS OF THE GREAT OCEAN LINERS IN HISTORIC PHOTOGRAPHS, William H. Miller, Jr. Some 200 superb photographs capture exquisite interiors of world's great "floating palaces"—1890's to 1980's: *Titanic, Ile de France, Queen Elizabeth, United States, Europa,* more. Approx. 200 black-and-white photographs. Captions. Text. Introduction. 160pp. 8⅞ × 11¾. 24756-2 Pa. $9.95

THE GREAT LUXURY LINERS, 1927–1954: A Photographic Record, William H. Miller, Jr. Nostalgic tribute to heyday of ocean liners. 186 photos of Ile de France, Normandie, Leviathan, Queen Elizabeth, United States, many others. Interior and exterior views. Introduction. Captions. 160pp. 9 × 12. 24056-8 Pa. $10.95

A NATURAL HISTORY OF THE DUCKS, John Charles Phillips. Great landmark of ornithology offers complete detailed coverage of nearly 200 species and subspecies of ducks: gadwall, sheldrake, merganser, pintail, many more. 74 full-color plates, 102 black-and-white. Bibliography. Total of 1,920pp. 8⅝ × 11¼. 25141-1, 25142-X Cloth. Two-vol. set $100.00

THE SEAWEED HANDBOOK: An Illustrated Guide to Seaweeds from North Carolina to Canada, Thomas F. Lee. Concise reference covers 78 species. Scientific and common names, habitat, distribution, more. Finding keys for easy identification. 224pp. 5⅜ × 8½. 25215-9 Pa. $6.95

THE TEN BOOKS OF ARCHITECTURE: The 1755 Leoni Edition, Leon Battista Alberti. Rare classic helped introduce the glories of ancient architecture to the Renaissance. 68 black-and-white plates. 336pp. 8⅜ × 11¼. 25239-6 Pa. $14.95

MISS MACKENZIE, Anthony Trollope. Minor masterpieces by Victorian master unmasks many truths about life in 19th-century England. First inexpensive edition in years. 392pp. 5⅜ × 8½. 25201-9 Pa. $8.95

THE RIME OF THE ANCIENT MARINER, Gustave Doré, Samuel Taylor Coleridge. Dramatic engravings considered by many to be his greatest work. The terrifying space of the open sea, the storms and whirlpools of an unknown ocean, the ice of Antarctica, more—all rendered in a powerful, chilling manner. Full text. 38 plates. 77pp. 9¼ × 12. 22305-1 Pa. $4.95

THE EXPEDITIONS OF ZEBULON MONTGOMERY PIKE, Zebulon Montgomery Pike. Fascinating first-hand accounts (1805–6) of exploration of Mississippi River, Indian wars, capture by Spanish dragoons, much more. 1,088pp. 5⅜ × 8½. 25254-X, 25255-8 Pa. Two-vol. set $25.90

CATALOG OF DOVER BOOKS

A CONCISE HISTORY OF PHOTOGRAPHY: Third Revised Edition, Helmut Gernsheim. Best one-volume history—camera obscura, photochemistry, daguerreotypes, evolution of cameras, film, more. Also artistic aspects—landscape, portraits, fine art, etc. 281 black-and-white photographs. 26 in color. 176pp. 8⅜ × 11¼. 25128-4 Pa. $13.95

THE DORÉ BIBLE ILLUSTRATIONS, Gustave Doré. 241 detailed plates from the Bible: the Creation scenes, Adam and Eve, Flood, Babylon, battle sequences, life of Jesus, etc. Each plate is accompanied by the verses from the King James version of the Bible. 241pp. 9 × 12. 23004-X Pa. $9.95

HUGGER-MUGGER IN THE LOUVRE, Elliot Paul. Second Homer Evans mystery-comedy. Theft at the Louvre involves sleuth in hilarious, madcap caper. "A knockout."—Books. 336pp. 5⅜ × 8½. 25185-3 Pa. $5.95

FLATLAND, E. A. Abbott. Intriguing and enormously popular science-fiction classic explores the complexities of trying to survive as a two-dimensional being in a three-dimensional world. Amusingly illustrated by the author. 16 illustrations. 103pp. 5⅜ × 8½. 20001-9 Pa. $2.50

THE HISTORY OF THE LEWIS AND CLARK EXPEDITION, Meriwether Lewis and William Clark, edited by Elliott Coues. Classic edition of Lewis and Clark's day-by-day journals that later became the basis for U.S. claims to Oregon and the West. Accurate and invaluable geographical, botanical, biological, meteorological and anthropological material. Total of 1,508pp. 5⅜ × 8½. 21268-8, 21269-6, 21270-X Pa. Three-vol. set $26.85

LANGUAGE, TRUTH AND LOGIC, Alfred J. Ayer. Famous, clear introduction to Vienna, Cambridge schools of Logical Positivism. Role of philosophy, elimination of metaphysics, nature of analysis, etc. 160pp. 5⅜ × 8½. (Available in U.S. and Canada only) 20010-8 Pa. $3.95

MATHEMATICS FOR THE NONMATHEMATICIAN, Morris Kline. Detailed, college-level treatment of mathematics in cultural and historical context, with numerous exercises. For liberal arts students. Preface. Recommended Reading Lists. Tables. Index. Numerous black-and-white figures. xvi + 641pp. 5⅜ × 8½. 24823-2 Pa. $11.95

HANDBOOK OF PICTORIAL SYMBOLS, Rudolph Modley. 3,250 signs and symbols, many systems in full; official or heavy commercial use. Arranged by subject. Most in Pictorial Archive series. 143pp. 8⅜ × 11. 23357-X Pa. $6.95

INCIDENTS OF TRAVEL IN YUCATAN, John L. Stephens. Classic (1843) exploration of jungles of Yucatan, looking for evidences of Maya civilization. Travel adventures, Mexican and Indian culture, etc. Total of 669pp. 5⅜ × 8½. 20926-1, 20927-X Pa., Two-vol. set $11.90

DEGAS: An Intimate Portrait, Ambroise Vollard. Charming, anecdotal memoir by famous art dealer of one of the greatest 19th-century French painters. 14 black-and-white illustrations. Introduction by Harold L. Van Doren. 96pp. 5⅜ × 8½.
25131-4 Pa. $4.95

PERSONAL NARRATIVE OF A PILGRIMAGE TO ALMANDINAH AND MECCAH, Richard Burton. Great travel classic by remarkably colorful personality. Burton, disguised as a Moroccan, visited sacred shrines of Islam, narrowly escaping death. 47 illustrations. 959pp. 5⅜ × 8½. 21217-3, 21218-1 Pa., Two-vol. set $19.90

PHRASE AND WORD ORIGINS, A. H. Holt. Entertaining, reliable, modern study of more than 1,200 colorful words, phrases, origins and histories. Much unexpected information. 254pp. 5⅜ × 8½. 20758-7 Pa. $5.95

THE RED THUMB MARK, R. Austin Freeman. In this first Dr. Thorndyke case, the great scientific detective draws fascinating conclusions from the nature of a single fingerprint. Exciting story, authentic science. 320pp. 5⅜ × 8½. (Available in U.S. only) 25210-8 Pa. $6.95

AN EGYPTIAN HIEROGLYPHIC DICTIONARY, E. A. Wallis Budge. Monumental work containing about 25,000 words or terms that occur in texts ranging from 3000 B.C. to 600 A.D. Each entry consists of a transliteration of the word, the word in hieroglyphs, and the meaning in English. 1,314pp. 6⅜ × 10.
23615-3, 23616-1 Pa., Two-vol. set $31.90

THE COMPLEAT STRATEGYST: Being a Primer on the Theory of Games of Strategy, J. D. Williams. Highly entertaining classic describes, with many illustrated examples, how to select best strategies in conflict situations. Prefaces. Appendices. xvi + 268pp. 5⅜ × 8½. 25101-2 Pa. $5.95

THE ROAD TO OZ, L. Frank Baum. Dorothy meets the Shaggy Man, little Button-Bright and the Rainbow's beautiful daughter in this delightful trip to the magical Land of Oz. 272pp. 5⅜ × 8. 25208-6 Pa. $5.95

POINT AND LINE TO PLANE, Wassily Kandinsky. Seminal exposition of role of point, line, other elements in non-objective painting. Essential to understanding 20th-century art. 127 illustrations. 192pp. 6½ × 9¼. 23808-3 Pa. $4.95

LADY ANNA, Anthony Trollope. Moving chronicle of Countess Lovel's bitter struggle to win for herself and daughter Anna their rightful rank and fortune—perhaps at cost of sanity itself. 384pp. 5⅜ × 8½. 24669-8 Pa. $8.95

EGYPTIAN MAGIC, E. A. Wallis Budge. Sums up all that is known about magic in Ancient Egypt: the role of magic in controlling the gods, powerful amulets that warded off evil spirits, scarabs of immortality, use of wax images, formulas and spells, the secret name, much more. 253pp. 5⅜ × 8½. 22681-6 Pa. $4.50

THE DANCE OF SIVA, Ananda Coomaraswamy. Preeminent authority unfolds the vast metaphysic of India: the revelation of her art, conception of the universe, social organization, etc. 27 reproductions of art masterpieces. 192pp. 5⅜ × 8½.
24817-8 Pa. $5.95

CHRISTMAS CUSTOMS AND TRADITIONS, Clement A. Miles. Origin, evolution, significance of religious, secular practices. Caroling, gifts, yule logs, much more. Full, scholarly yet fascinating; non-sectarian. 400pp. 5⅜ × 8½.
23354-5 Pa. $6.95

THE HUMAN FIGURE IN MOTION, Eadweard Muybridge. More than 4,500 stopped-action photos, in action series, showing undraped men, women, children jumping, lying down, throwing, sitting, wrestling, carrying, etc. 390pp. 7⅞ × 10⅝.
20204-6 Cloth. $21.95

THE MAN WHO WAS THURSDAY, Gilbert Keith Chesterton. Witty, fast-paced novel about a club of anarchists in turn-of-the-century London. Brilliant social, religious, philosophical speculations. 128pp. 5⅜ × 8½. 25121-7 Pa. $3.95

A CEZANNE SKETCHBOOK: Figures, Portraits, Landscapes and Still Lifes, Paul Cezanne. Great artist experiments with tonal effects, light, mass, other qualities in over 100 drawings. A revealing view of developing master painter, precursor of Cubism. 102 black-and-white illustrations. 144pp. 8¾ × 6⅜. 24790-2 Pa. $5.95

AN ENCYCLOPEDIA OF BATTLES: Accounts of Over 1,560 Battles from 1479 B.C. to the Present, David Eggenberger. Presents essential details of every major battle in recorded history, from the first battle of Megiddo in 1479 B.C. to Grenada in 1984. List of Battle Maps. New Appendix covering the years 1967–1984. Index. 99 illustrations. 544pp. 6½ × 9¼. 24913-1 Pa. $14.95

AN ETYMOLOGICAL DICTIONARY OF MODERN ENGLISH, Ernest Weekley. Richest, fullest work, by foremost British lexicographer. Detailed word histories. Inexhaustible. Total of 856pp. 6½ × 9¼.
21873-2, 21874-0 Pa., Two-vol. set $17.00

WEBSTER'S AMERICAN MILITARY BIOGRAPHIES, edited by Robert McHenry. Over 1,000 figures who shaped 3 centuries of American military history. Detailed biographies of Nathan Hale, Douglas MacArthur, Mary Hallaren, others. Chronologies of engagements, more. Introduction. Addenda. 1,033 entries in alphabetical order. xi + 548pp. 6½ × 9¼. (Available in U.S. only)
24758-9 Pa. $13.95

LIFE IN ANCIENT EGYPT, Adolf Erman. Detailed older account, with much not in more recent books: domestic life, religion, magic, medicine, commerce, and whatever else needed for complete picture. Many illustrations. 597pp. 5⅜ × 8½.
22632-8 Pa. $8.95

HISTORIC COSTUME IN PICTURES, Braun & Schneider. Over 1,450 costumed figures shown, covering a wide variety of peoples: kings, emperors, nobles, priests, servants, soldiers, scholars, townsfolk, peasants, merchants, courtiers, cavaliers, and more. 256pp. 8⅜ × 11¼. 23150-X Pa. $9.95

THE NOTEBOOKS OF LEONARDO DA VINCI, edited by J. P. Richter. Extracts from manuscripts reveal great genius; on painting, sculpture, anatomy, sciences, geography, etc. Both Italian and English. 186 ms. pages reproduced, plus 500 additional drawings, including studies for *Last Supper, Sforza* monument, etc. 860pp. 7⅞ × 10¾. (Available in U.S. only) 22572-0, 22573-9 Pa., Two-vol. set $31.90

THE ART NOUVEAU STYLE BOOK OF ALPHONSE MUCHA: All 72 Plates from "Documents Decoratifs" in Original Color, Alphonse Mucha. Rare copyright-free design portfolio by high priest of Art Nouveau. Jewelry, wallpaper, stained glass, furniture, figure studies, plant and animal motifs, etc. Only complete one-volume edition. 80pp. 9⅜ × 12¼. 24044-4 Pa. $9.95

ANIMALS: 1,419 COPYRIGHT-FREE ILLUSTRATIONS OF MAMMALS, BIRDS, FISH, INSECTS, ETC., edited by Jim Harter. Clear wood engravings present, in extremely lifelike poses, over 1,000 species of animals. One of the most extensive pictorial sourcebooks of its kind. Captions. Index. 284pp. 9 × 12. 23766-4 Pa. $9.95

OBELISTS FLY HIGH, C. Daly King. Masterpiece of American detective fiction, long out of print, involves murder on a 1935 transcontinental flight—"a very thrilling story"—NY Times. Unabridged and unaltered republication of the edition published by William Collins Sons & Co. Ltd., London, 1935. 288pp. 5⅜ × 8½. (Available in U.S. only) 25036-9 Pa. $5.95

VICTORIAN AND EDWARDIAN FASHION: A Photographic Survey, Alison Gernsheim. First fashion history completely illustrated by contemporary photographs. Full text plus 235 photos, 1840-1914, in which many celebrities appear. 240pp. 6½ × 9¼. 24205-6 Pa. $6.95

THE ART OF THE FRENCH ILLUSTRATED BOOK, 1700-1914, Gordon N. Ray. Over 630 superb book illustrations by Fragonard, Delacroix, Daumier, Doré, Grandville, Manet, Mucha, Steinlen, Toulouse-Lautrec and many others. Preface. Introduction. 633 halftones. Indices of artists, authors & titles, binders and provenances. Appendices. Bibliography. 608pp. 8⅜ × 11¼. 25086-5 Pa. $24.95

THE WONDERFUL WIZARD OF OZ, L. Frank Baum. Facsimile in full color of America's finest children's classic. 143 illustrations by W. W. Denslow. 267pp. 5⅜ × 8½. 20691-2 Pa. $7.95

FRONTIERS OF MODERN PHYSICS: New Perspectives on Cosmology, Relativity, Black Holes and Extraterrestrial Intelligence, Tony Rothman, et al. For the intelligent layman. Subjects include: cosmological models of the universe; black holes; the neutrino; the search for extraterrestrial intelligence. Introduction. 46 black-and-white illustrations. 192pp. 5⅜ × 8½. 24587-X Pa. $7.95

THE FRIENDLY STARS, Martha Evans Martin & Donald Howard Menzel. Classic text marshalls the stars together in an engaging, non-technical survey, presenting them as sources of beauty in night sky. 23 illustrations. Foreword. 2 star charts. Index. 147pp. 5⅜ × 8½. 21099-5 Pa. $3.95

FADS AND FALLACIES IN THE NAME OF SCIENCE, Martin Gardner. Fair, witty appraisal of cranks, quacks, and quackeries of science and pseudoscience: hollow earth, Velikovsky, orgone energy, Dianetics, flying saucers, Bridey Murphy, food and medical fads, etc. Revised, expanded In the Name of Science. "A very able and even-tempered presentation."—The New Yorker. 363pp. 5⅜ × 8.

20394-8 Pa. $6.95

ANCIENT EGYPT: ITS CULTURE AND HISTORY, J. E Manchip White. From pre-dynastics through Ptolemies: society, history, political structure, religion, daily life, literature, cultural heritage. 48 plates. 217pp. 5⅜ × 8½. 22548-8 Pa. $5.95

SIR HARRY HOTSPUR OF HUMBLETHWAITE, Anthony Trollope. Incisive, unconventional psychological study of a conflict between a wealthy baronet, his idealistic daughter, and their scapegrace cousin. The 1870 novel in its first inexpensive edition in years. 250pp. 5⅜ × 8½. 24953-0 Pa. $5.95

LASERS AND HOLOGRAPHY, Winston E. Kock. Sound introduction to burgeoning field, expanded (1981) for second edition. Wave patterns, coherence, lasers, diffraction, zone plates, properties of holograms, recent advances. 84 illustrations. 160pp. 5⅜ × 8¼. (Except in United Kingdom) 24041-X Pa. $3.95

INTRODUCTION TO ARTIFICIAL INTELLIGENCE: SECOND, EN-LARGED EDITION, Philip C. Jackson, Jr. Comprehensive survey of artificial intelligence—the study of how machines (computers) can be made to act intelligently. Includes introductory and advanced material. Extensive notes updating the main text. 132 black-and-white illustrations. 512pp. 5⅜ × 8½. 24864-X Pa. $8.95

HISTORY OF INDIAN AND INDONESIAN ART, Ananda K. Coomaraswamy. Over 400 illustrations illuminate classic study of Indian art from earliest Harappa finds to early 20th century. Provides philosophical, religious and social insights. 304pp. 6⅜ × 9⅜. 25005-9 Pa. $9.95

THE GOLEM, Gustav Meyrink. Most famous supernatural novel in modern European literature, set in Ghetto of Old Prague around 1890. Compelling story of mystical experiences, strange transformations, profound terror. 13 black-and-white illustrations. 224pp. 5⅜ × 8½. (Available in U.S. only) 25025-3 Pa. $6.95

ARMADALE, Wilkie Collins. Third great mystery novel by the author of *The Woman in White* and *The Moonstone*. Original magazine version with 40 illustrations. 597pp. 5⅜ × 8½. 23429-0 Pa. $9.95

PICTORIAL ENCYCLOPEDIA OF HISTORIC ARCHITECTURAL PLANS, DETAILS AND ELEMENTS: With 1,880 Line Drawings of Arches, Domes, Doorways, Facades, Gables, Windows, etc., John Theodore Haneman. Sourcebook of inspiration for architects, designers, others. Bibliography. Captions. 141pp. 9 × 12. 24605-1 Pa. $7.95

BENCHLEY LOST AND FOUND, Robert Benchley. Finest humor from early 30's, about pet peeves, child psychologists, post office and others. Mostly unavailable elsewhere. 73 illustrations by Peter Arno and others. 183pp. 5⅜ × 8½. 22410-4 Pa. $4.95

ERTÉ GRAPHICS, Erté. Collection of striking color graphics: *Seasons, Alphabet, Numerals, Aces* and *Precious Stones*. 50 plates, including 4 on covers. 48pp. 9⅜ × 12¼. 23580-7 Pa. $6.95

THE JOURNAL OF HENRY D. THOREAU, edited by Bradford Torrey, F. H. Allen. Complete reprinting of 14 volumes, 1837–61, over two million words; the sourcebooks for *Walden*, etc. Definitive. All original sketches, plus 75 photographs. 1,804pp. 8½ × 12¼. 20312-3, 20313-1 Cloth., Two-vol. set $120.00

CASTLES: THEIR CONSTRUCTION AND HISTORY, Sidney Toy. Traces castle development from ancient roots. Nearly 200 photographs and drawings illustrate moats, keeps, baileys, many other features. Caernarvon, Dover Castles, Hadrian's Wall, Tower of London, dozens more. 256pp. 5⅜ × 8¼. 24898-4 Pa. $6.95

AMERICAN CLIPPER SHIPS: 1833–1858, Octavius T. Howe & Frederick C. Matthews. Fully-illustrated, encyclopedic review of 352 clipper ships from the period of America's greatest maritime supremacy. Introduction. 109 halftones. 5 black-and-white line illustrations. Index. Total of 928pp. 5⅜ × 8½.
25115-2, 25116-0 Pa., Two-vol. set $17.90

TOWARDS A NEW ARCHITECTURE, Le Corbusier. Pioneering manifesto by great architect, near legendary founder of "International School." Technical and aesthetic theories, views on industry, economics, relation of form to function, "mass-production spirit," much more. Profusely illustrated. Unabridged translation of 13th French edition. Introduction by Frederick Etchells. 320pp. 6⅛ × 9¼. (Available in U.S. only)
25023-7 Pa. $8.95

THE BOOK OF KELLS, edited by Blanche Cirker. Inexpensive collection of 32 full-color, full-page plates from the greatest illuminated manuscript of the Middle Ages, painstakingly reproduced from rare facsimile edition. Publisher's Note. Captions. 32pp. 9⅜ × 12¼.
24345-1 Pa. $4.95

BEST SCIENCE FICTION STORIES OF H. G. WELLS, H. G. Wells. Full novel *The Invisible Man*, plus 17 short stories: "The Crystal Egg," "Aepyornis Island," "The Strange Orchid," etc. 303pp. 5⅜ × 8½. (Available in U.S. only)
21531-8 Pa. $6.95

AMERICAN SAILING SHIPS: Their Plans and History, Charles G. Davis. Photos, construction details of schooners, frigates, clippers, other sailcraft of 18th to early 20th centuries—plus entertaining discourse on design, rigging, nautical lore, much more. 137 black-and-white illustrations. 240pp. 6⅛ × 9¼.
24658-2 Pa. $6.95

ENTERTAINING MATHEMATICAL PUZZLES, Martin Gardner. Selection of author's favorite conundrums involving arithmetic, money, speed, etc., with lively commentary. Complete solutions. 112pp. 5⅜ × 8½.
25211-6 Pa. $2.95

THE WILL TO BELIEVE, HUMAN IMMORTALITY, William James. Two books bound together. Effect of irrational on logical, and arguments for human immortality. 402pp. 5⅜ × 8½.
20291-7 Pa. $7.95

THE HAUNTED MONASTERY and THE CHINESE MAZE MURDERS, Robert Van Gulik. 2 full novels by Van Gulik continue adventures of Judge Dee and his companions. An evil Taoist monastery, seemingly supernatural events; overgrown topiary maze that hides strange crimes. Set in 7th-century China. 27 illustrations. 328pp. 5⅜ × 8½.
23502-5 Pa. $6.95

CELEBRATED CASES OF JUDGE DEE (DEE GOONG AN), translated by Robert Van Gulik. Authentic 18th-century Chinese detective novel; Dee and associates solve three interlocked cases. Led to Van Gulik's own stories with same characters. Extensive introduction. 9 illustrations. 237pp. 5⅜ × 8½.
23337-5 Pa. $4.95

Prices subject to change without notice.

Available at your book dealer or write for free catalog to Dept. GI, Dover Publications, Inc., 31 East 2nd St., Mineola, N.Y. 11501. Dover publishes more than 175 books each year on science, elementary and advanced mathematics, biology, music, art, literary history, social sciences and other areas.